SQA

2017 SQA Specimen and Past Papers with Answers

National 5
CHEMISTRY

2016 & 2017 Exams
and 2017 Specimen Question Paper

HODDER
GIBSON
AN HACHETTE UK COMPANY

This book contains the official SQA 2016 and 2017 Exams, and the 2017 Specimen Question Paper for National 5 Chemistry, with associated SQA-approved answers modified from the official marking instructions that accompany the paper.

In addition the book contains study skills advice. This has been specially commissioned by Hodder Gibson, and has been written by experienced senior teachers and examiners in line with the new National 5 syllabus and assessment outlines. This is not SQA material but has been devised to provide further guidance for National 5 examinations.

Hodder Gibson is grateful to the copyright holders, as credited on the final page of the Answer section, for permission to use their material. Every effort has been made to trace the copyright holders and to obtain their permission for the use of copyright material. Hodder Gibson will be happy to receive information allowing us to rectify any error or omission in future editions.

Hachette UK's policy is to use papers that are natural, renewable and recyclable products and made from wood grown in sustainable forests. The logging and manufacturing processes are expected to conform to the environmental regulations of the country of origin.

Orders: please contact Bookpoint Ltd, 130 Park Drive, Milton Park, Abingdon, Oxon OX14 4SE. Telephone: (44) 01235 827720. Fax: (44) 01235 400454. Lines are open 9.00–5.00, Monday to Saturday, with a 24-hour message answering service. Visit our website at www.hoddereducation.co.uk. Hodder Gibson can be contacted direct on: Tel: 0141 333 4650; Fax: 0141 404 8188; email: hoddergibson@hodder.co.uk

This collection first published in 2017 by
Hodder Gibson, an imprint of Hodder Education,
An Hachette UK Company
211 St Vincent Street
Glasgow G2 5QY

Typeset by Aptara, Inc.

Printed in the UK

A catalogue record for this title is available from the British Library

ISBN: 978-1-5104-2159-2

2 1

2018 2017

MIX
Paper from
responsible sources
FSC™ C104740

Introduction

National 5 Chemistry

This book of SQA Past Papers contains the actual 2016 and 2017 Question Papers (with answers at the back of the book). As the course specifications have changed, due to the removal of Unit Assessments and the 'strengthening' of the exams, a new Specimen Question Paper, which reflects the revised exam structure, is also included. The Specimen Question Paper reflects the content and duration of the exam in 2018. All of the question papers included in the book (2016, 2017 and the SQP) provide excellent practice for the final exams.

The 2016 and 2017 exams still provide excellent representative exam practice. Using them as part of your revision will help you to learn the vital skills and techniques needed for the exam, and will help you to identify any knowledge gaps you may have, prior to the exam season in May–June.

The course

National 5 courses have changed.

Unit Assessments have been removed from all National 5 courses. This means that you do not have to pass Unit Assessments in order to be eligible to take the exam. It also means that the exams for 2018 onwards have been updated and strengthened to assess your knowledge and skills across the whole course.

To achieve a pass in National 5 Chemistry there are two main components.

Component 1 – Assignment

You are required to submit an assignment that is worth 20% (20 marks) of your final grade. This assignment will be based on research but it is important that you are aware that practical/experimental/field work is a mandatory feature of your assignment so has to be included. This assignment requires you to apply skills, knowledge and understanding to investigate a relevant topic in chemistry and its effect on the environment and/or society. Your school or college will provide you with a Candidate's Guide for this assignment, which has been produced by the SQA. This guide gives guidance on what is required to complete the report and gain as many marks as possible. When you have completed your research and collected your experimental data, you will be allocated a maximum of 1 hour 30 minutes to complete your report in school.

Your assignment report will be marked by the SQA.

Component 2 – The Question Paper

The question paper will assess breadth and depth of knowledge and understanding from across the whole course. It is worth 80% of your final grade and is marked out of 100 possible marks. The question paper will require you to:

- Make statements, provide explanations, and describe information to demonstrate knowledge and understanding.
- Apply knowledge and understanding to new situations to solve problems.
- Plan and design experiments.
- Present information in various forms such as graphs, tables, etc.
- Perform calculations based on information given.
- Give predictions or make generalisations based on information given.
- Draw conclusions based on information given.
- Suggest improvement to experiments to improve the accuracy of results obtained or to improve safety.

To achieve a "C" grade in National 5 Chemistry you must achieve about 50% of the 120 marks available when the two components, i.e. the Question Paper and the Assignment are combined. For a B you will need 60%, while for an "A" grade you must ensure that you gain as many of the available marks as possible, and at least 70%.

Each SQA Past Paper consists of two sections. (A marking scheme for each section is provided at the end of this book.)

- Section 1 will contain objective questions (multiple choice) and will have 25 marks.
- Section 2 will contain restricted and extended response questions and will have 75 marks.

Each SQA Past Paper contains a variety of questions including some that require:

- demonstration and application of knowledge, and understanding of the mandatory content of the course from across the three areas of the course: chemical changes and structure; nature's chemistry; and chemistry in society.
- application of scientific inquiry skills.

How to use this book

This book can be used in two ways:

1. You can complete an entire paper under exam conditions, without the use of books or notes, and then mark the papers using the marking scheme provided. This method gives you a clear indication of the level you are working at and should highlight the content areas that you need to work on before attempting the next paper. This method also allows you to see your progress as you complete each paper.

2. You can complete a paper using your notes and books. Try the question first and then refer to your notes if you are unable to answer the question. This is a form of studying and by doing this you will cover all the areas of content that you are weakest in. You should notice that you are referring to your notes less with each paper completed.

Try to practise as many questions as possible. This will get you used to the language used in the papers and ultimately improve your chances of success.

Some hints and tips

Below is a list of hints and tips that will help you to achieve your full potential in the National 5 exam.

- Ensure that you **read each question carefully**. Scanning the question and missing the main points results in mistakes being made. Some students highlight the main points of a question with a highlighter pen to ensure that they don't miss anything out.

- Open ended questions include the statement **"Using your knowledge of chemistry"**. These questions provide you with an opportunity to show off your chemistry knowledge. To obtain the three marks on offer for these questions, you must demonstrate a good understanding of the chemistry involved and provide a logically correct answer to the question posed.

- When doing calculations, ensure that you **show all of your working**. If you make a simple arithmetical mistake you may still be awarded some of the marks, but only if your working is laid out clearly so that the examiner can see where you went wrong and what you did correctly. Just giving the answers is very risky so you should always show your working.

- **Attempt all questions.** Giving no answer at all means that you will definitely not gain any marks.

- When you are required to read a passage to answer a question, ensure that you **read it carefully** as the information you require is contained within it. It may not be obvious at first, but the answers will be contained within the passage.

- If you are asked to "explain" in a question, then you must **explain your answer fully**. For example, if you are asked to explain how a covalent bond holds atoms together then you cannot simply say:

    "A covalent bond is a shared pair of electrons between atoms in a non-metal."

    This answer tells the examiner what a covalent bond is, but does not explain how it holds the atoms together. To gain the marks, an answer similar to this should be written:

    "A covalent bond is a shared pair of electrons between atoms in a non-metal. The shared electrons are attracted to the nuclei of both atoms, which creates a tug-of-war effect creating the covalent bond."

- You may be required to draw one graph in each exam. To obtain all the marks, ensure that the graphs have **labels, units, points plotted correctly** and a line of "best fit" drawn between the points.

- Use your **data booklet** when you are asked to write formulas, ionic formulas, formula mass etc. You have the data booklet in front of you so use it to double check the numbers you require.

- Work on your **timing**. The multiple-choice section (Section 1) should take approximately 40 minutes. Attempt to answer the multiple-choice questions before you look at the four possible answers, as this will improve your confidence. Use scrap paper when required to scribble down structural formulae, calculations, chemical formulae etc., as this will reduce your chance of making errors. If you are finding the question difficult, try to eliminate the obviously wrong answers to increase your chances.

- When asked to **predict or estimate** based on information from a graph or a table, then take your time to look for patterns. For example, if asked to predict a boiling point, try to establish if there is a regular change in boiling point and use that regular pattern to establish the unknown boiling point.

- When drawing a **diagram** of an experiment ask yourself the question, "Would this work if I set it up exactly like this in the lab?" Ensure that the method you have drawn would produce the desired results <u>safely</u>. If, for example, you are heating a flammable reactant such as alcohol then you will not gain the marks if you heat it with a Bunsen burner in your diagram; a water bath would be much safer! Make sure your diagram is labelled clearly.

Good luck!

Remember that the rewards for passing National 5 Chemistry are well worth it! Your pass will help you get the future you want for yourself. In the exam, be confident in your own ability. If you're not sure how to answer a question, trust your instincts and just give it a go anyway. Keep calm and don't panic! GOOD LUCK!

Study Skills – what you need to know to pass exams!

Pause for thought

Many students might skip quickly through a page like this. After all, we all know how to revise. Do you really though?

Think about this:

"IF YOU ALWAYS DO WHAT YOU ALWAYS DO, YOU WILL ALWAYS GET WHAT YOU HAVE ALWAYS GOT."

Do you like the grades you get? Do you want to do better? If you get full marks in your assessment, then that's great! Change nothing! This section is just to help you get that little bit better than you already are.

There are two main parts to the advice on offer here. The first part highlights fairly obvious things but which are also very important. The second part makes suggestions about revision that you might not have thought about but which WILL help you.

Part 1

DOH! It's so obvious but …

Start revising in good time

Don't leave it until the last minute – this will make you panic.

Make a revision timetable that sets out work time AND play time.

Sleep and eat!

Obvious really, and very helpful. Avoid arguments or stressful things too – even games that wind you up. You need to be fit, awake and focused!

Know your place!

Make sure you know exactly **WHEN and WHERE** your exams are.

Know your enemy!

Make sure you know what to expect in the exam.

How is the paper structured?

How much time is there for each question?

What types of question are involved?

Which topics seem to come up time and time again?

Which topics are your strongest and which are your weakest?

Are all topics compulsory or are there choices?

Learn by DOING!

There is no substitute for past papers and practice papers – they are simply essential! Tackling this collection of papers and answers is exactly the right thing to be doing as your exams approach.

Part 2

People learn in different ways. Some like low light, some bright. Some like early morning, some like evening / night. Some prefer warm, some prefer cold. But everyone uses their BRAIN and the brain works when it is active. Passive learning – sitting gazing at notes – is the most INEFFICIENT way to learn anything. Below you will find tips and ideas for making your revision more effective and maybe even more enjoyable. What follows gets your brain active, and active learning works!

Activity 1 – Stop and review

Step 1

When you have done no more than 5 minutes of revision reading STOP!

Step 2

Write a heading in your own words which sums up the topic you have been revising.

Step 3

Write a summary of what you have revised in no more than two sentences. Don't fool yourself by saying, "I know it, but I cannot put it into words". That just means you don't know it well enough. If you cannot write your summary, revise that section again, knowing that you must write a summary at the end of it. Many of you will have notebooks full of blue/black ink writing. Many of the pages will not be especially attractive or memorable so try to liven them up a bit with colour as you are reviewing and rewriting. **This is a great memory aid, and memory is the most important thing.**

Activity 2 – Use technology!

Why should everything be written down? Have you thought about "mental" maps, diagrams, cartoons and colour to help you learn? And rather than write down notes, why not record your revision material?

What about having a text message revision session with friends? Keep in touch with them to find out how and what they are revising and share ideas and questions.

Why not make a video diary where you tell the camera what you are doing, what you think you have learned and what you still have to do? No one has to see or hear it, but the process of having to organise your thoughts in a formal way to explain something is a very important learning practice.

Be sure to make use of electronic files. You could begin to summarise your class notes. Your typing might be slow, but it will get faster and the typed notes will be easier to read than the scribbles in your class notes. Try to add different fonts and colours to make your work stand out. You can easily Google relevant pictures, cartoons and diagrams which you can copy and paste to make your work more attractive and **MEMORABLE**.

Activity 3 – This is it. Do this and you will know lots!

Step 1

In this task you must be very honest with yourself! Find the SQA syllabus for your subject (www.sqa.org.uk). Look at how it is broken down into main topics called MANDATORY knowledge. That means stuff you MUST know.

Step 2

BEFORE you do ANY revision on this topic, write a list of everything that you already know about the subject. It might be quite a long list but you only need to write it once. It shows you all the information that is already in your long-term memory so you know what parts you do not need to revise!

Step 3

Pick a chapter or section from your book or revision notes. Choose a fairly large section or a whole chapter to get the most out of this activity.

With a buddy, use Skype, Facetime, Twitter or any other communication you have, to play the game "If this is the answer, what is the question?". For example, if you are revising Geography and the answer you provide is "meander", your buddy would have to make up a question like "What is the word that describes a feature of a river where it flows slowly and bends often from side to side?".

Make up 10 "answers" based on the content of the chapter or section you are using. Give this to your buddy to solve while you solve theirs.

Step 4

Construct a wordsearch of at least 10 × 10 squares. You can make it as big as you like but keep it realistic. Work together with a group of friends. Many apps allow you to make wordsearch puzzles online. The words and phrases can go in any direction and phrases can be split. Your puzzle must only contain facts linked to the topic you are revising. Your task is to find 10 bits of information to hide in your puzzle, but you must not repeat information that you used in Step 3. DO NOT show where the words are. Fill up empty squares with random letters. Remember to keep a note of where your answers are hidden but do not show your friends. When you have a complete puzzle, exchange it with a friend to solve each other's puzzle.

Step 5

Now make up 10 questions (not "answers" this time) based on the same chapter used in the previous two tasks. Again, you must find NEW information that you have not yet used. Now it's getting hard to find that new information! Again, give your questions to a friend to answer.

Step 6

As you have been doing the puzzles, your brain has been actively searching for new information. Now write a NEW LIST that contains only the new information you have discovered when doing the puzzles. Your new list is the one to look at repeatedly for short bursts over the next few days. Try to remember more and more of it without looking at it. After a few days, you should be able to add words from your second list to your first list as you increase the information in your long-term memory.

FINALLY! Be inspired...

Make a list of different revision ideas and beside each one write **THINGS I HAVE** tried, **THINGS I WILL** try and **THINGS I MIGHT** try. Don't be scared of trying something new.

And remember – "FAIL TO PREPARE AND PREPARE TO FAIL!"

NATIONAL 5

2016

National Qualifications 2016

X713/75/02

Chemistry
Section 1—Questions

WEDNESDAY, 18 MAY

1:00 PM – 3:00 PM

Instructions for the completion of Section 1 are given on *Page two* of your question and answer booklet X713/75/01.

Record your answers on the answer grid on *Page three* of your question and answer booklet.

Necessary data will be found in the Chemistry Data Booklet for National 5.

Before leaving the examination room you must give your question and answer booklet to the Invigilator; if you do not, you may lose all the marks for this paper.

SECTION 1

1. When solid sodium chloride dissolves in water, a solution containing sodium ions and chloride ions is formed.

   Which of the following equations correctly shows the state symbols for this process?

   A $NaCl(s)$ + $H_2O(\ell)$ ⟶ $Na^+(\ell)$ + $Cl^-(\ell)$

   B $NaCl(s)$ + $H_2O(aq)$ ⟶ $Na^+(aq)$ + $Cl^-(aq)$

   C $NaCl(aq)$ + $H_2O(\ell)$ ⟶ $Na^+(aq)$ + $Cl^-(aq)$

   (D) $NaCl(s)$ + $H_2O(\ell)$ ⟶ $Na^+(aq)$ + $Cl^-(aq)$

2. The table shows the times taken for 0·5 g of magnesium to react completely with acid under different conditions.

Acid concentration (mol l^{-1})	Temperature (°C)	Reaction time (s)
0·1	20	80
0·1	25	60
0·2	30	20
0·2	40	10

   The time for 0·5 g of magnesium to react completely with 0·2 mol l^{-1} acid at 25 °C will be

   A   less than 10 s

   B   between 10 s and 20 s

   C   between 20 s and 60 s

   D   more than 80 s.

3. When an atom **X** of an element in Group 1 reacts to become X^+

   A   the mass number of **X** decreases

   B   the atomic number of **X** increases

   C   the charge of the nucleus increases

   D   the number of occupied energy levels decreases.

4. Which of the following does **not** contain covalent bonds?

   A   Sulfur

   B   Copper

   C   Oxygen

   (D) Hydrogen

5. Which of the following structures is **never** found in compounds?

    A   Ionic

    B   Monatomic

    C   Covalent network

    D   Covalent molecular

6. Which line in the table shows the properties of an ionic substance?

	Melting point (°C)	Boiling point (°C)	Conducts electricity	
			Solid	Liquid
A	19	80	no	no
B	655	1425	no	no
C	1450	1740	no	yes
D	1495	2927	yes	yes

7. What is the name of the compound with the formula Ag_2O?

    A   Silver(I) oxide

    B   Silver(II) oxide

    C   Silver(III) oxide

    D   Silver(IV) oxide

8. An element was burned in air. The product was added to water, producing a solution with a pH less than 7. The element could be

    A   tin

    B   zinc

    C   sulfur

    D   sodium.

9. When methane burns in a plentiful supply of air, the products are

    A   carbon and water

    B   carbon dioxide and water

    C   carbon monoxide and water

    D   carbon dioxide and hydrogen.

[Turn over

10. Which of the following compounds belongs to the same homologous series as the compound with the molecular formula C_3H_8?

A
```
      H   H
      |   |
  H — C — C — H
      |   |
  H — C — C — H
      |   |
      H   H
```

B
```
      H           H
      |           |
  H — C — C = C — C — H
      |   |   |   |
      H   H   H   H
```

C
```
              H
              |
          H — C — H
              |
      H   H   |   H
      |   |   |   |
  H — C — C — C — C — H
      |   |   |   |
      H   H   H   H
```

D
```
              H
              |
          H — C — H
              |
      H   H   |
      |   |   |
  H — C — C — C = C — H
      |   |   |   |
      H   H   H   H
```

11. 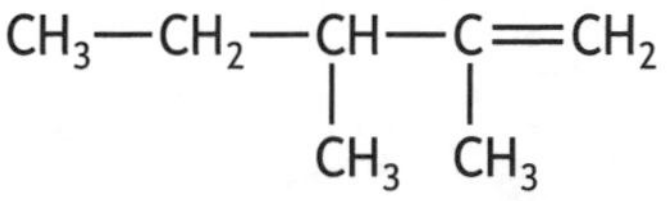 $CH_3-CH_2-CH-C=CH_2$
                              | |
                             CH_3 CH_3

The systematic name for the structure shown is

A   1,2-dimethylpent-1-ene

B   2,3-dimethylpent-1-ene

C   3,4-dimethylpent-4-ene

D   3,4-dimethylpent-1-ene.

12.   Two isomers of butene are

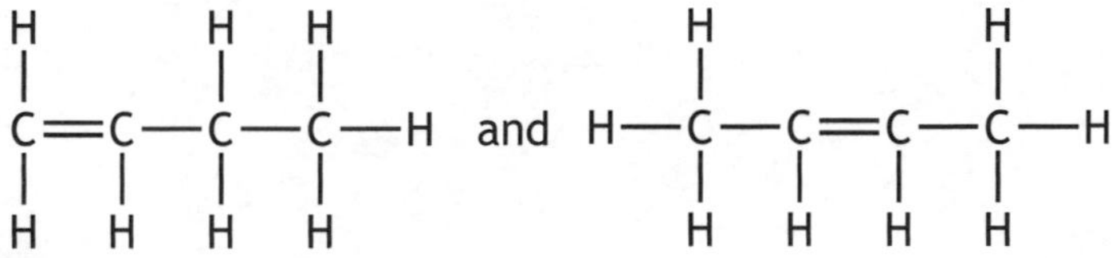

Which of the following structures represents a third isomer of butene?

A

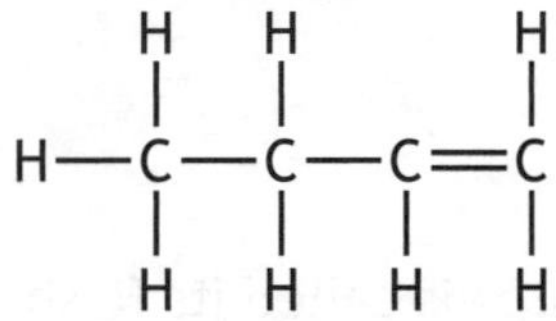

B

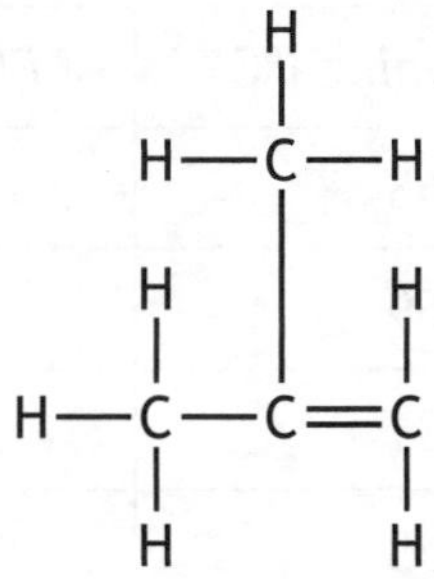

C

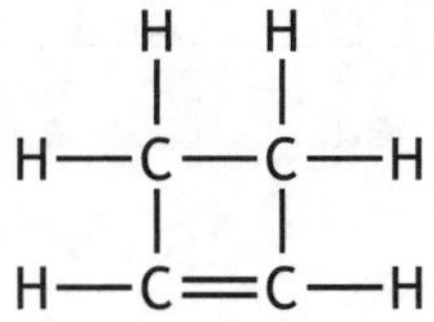

D

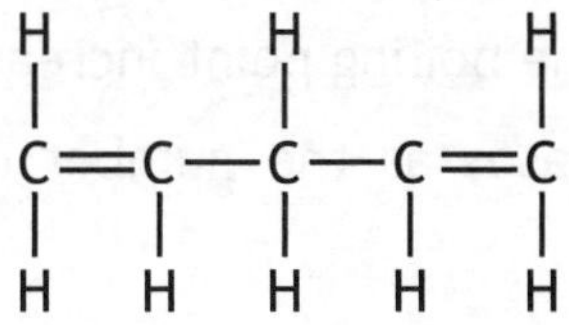

[Turn over

13. Which of the following structures represents an ester?

A

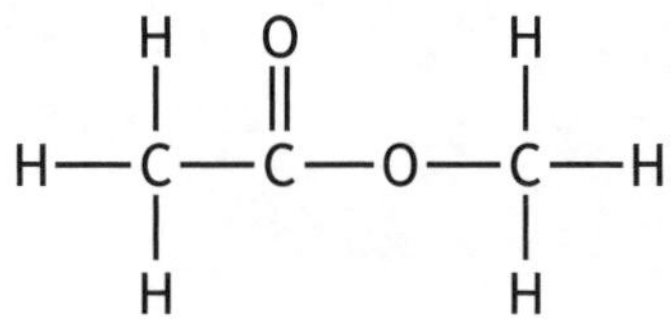

B

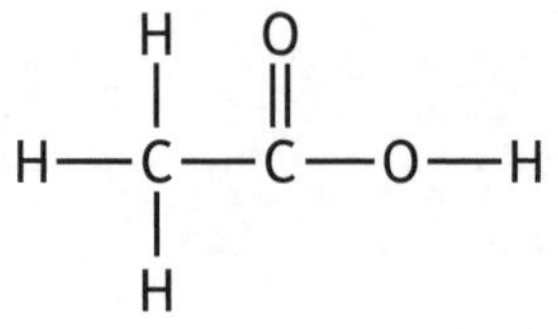

C

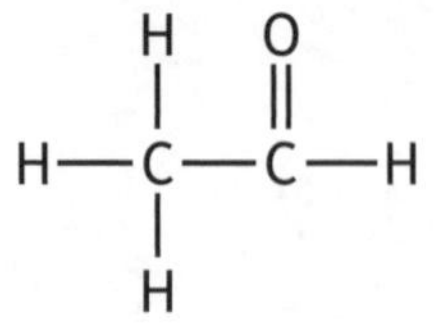

D

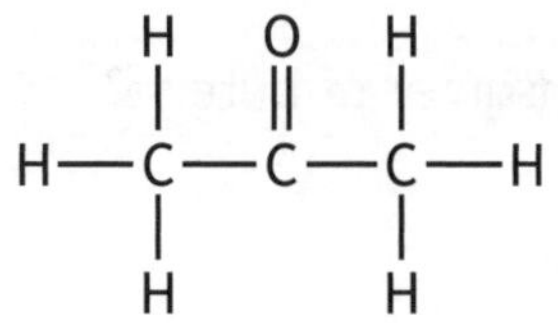

14. The lowest temperature at which a hydrocarbon ignites is called its flash point.

Hydrocarbon	Formula	Boiling point (°C)	Flash point (°C)
hexene	C_6H_{12}	63	−25
hexane	C_6H_{14}	69	−23
cyclohexane	C_6H_{12}	81	−20
heptane	C_7H_{16}	98	−1
octane	C_8H_{18}	126	15

Using information in the table, identify the correct statement.

A   Octane will ignite at 0 °C.

B   Hydrocarbons with the same molecular mass have the same flash point.

C   The flash point of a hydrocarbon increases as the boiling point increases.

D   In a homologous series the flash point decreases as the number of carbon atoms increases.

15. Which of the following metals can be obtained from its ore by heating with carbon monoxide?

    You may wish to use the data booklet to help you.

    A   Magnesium

    B   Aluminium

    C   Calcium

    D   Nickel

16. Polyesters are always made from monomers

    A   which are the same

    B   which are unsaturated

    C   with one functional group per molecule

    D   with two functional groups per molecule.

17. Some smoke detectors make use of radiation which is very easily stopped by tiny smoke particles moving between the radioactive source and the detector.

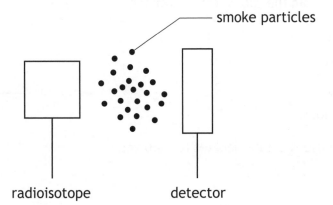

    radioisotope   detector

    The most suitable type of radioisotope for a smoke detector would be

    A   an alpha-emitter with a long half-life

    B   a gamma-emitter with a short half-life

    C   an alpha-emitter with a short half-life

    D   a gamma-emitter with a long half-life.

[Turn over for next question

18. Which particle will be formed when an atom of $^{234}_{90}$Th emits a β-particle?

  A $^{234}_{91}$Pa

  B $^{230}_{88}$Ra

  C $^{234}_{89}$Ac

  D $^{238}_{92}$U

19. ^{14}C has a half life of 5600 years. An analysis of charcoal from a wood fire shows that its ^{14}C content is 25 % of that in living wood.

  How many years have passed since the wood for the fire was cut?

  A 1400

  B 4200

  C 11 200

  D 16 800

20. A solution of potassium carbonate, made up using tap water, was found to be cloudy. This could result from the tap water containing

  A lithium ions

  B calcium ions

  C sodium ions

  D ammonium ions.

  You may wish to use the data booklet to help you.

[END OF SECTION 1. NOW ATTEMPT THE QUESTIONS IN SECTION 2 OF YOUR QUESTION AND ANSWER BOOKLET]

FOR OFFICIAL USE

National
Qualifications
2016

Mark

X713/75/01

Chemistry
Section 1—Answer Grid
And Section 2

WEDNESDAY, 18 MAY

1:00 PM — 3:00 PM

Fill in these boxes and read what is printed below.

Full name of centre

Town

Forename(s)

Surname

Number of seat

Date of birth

Day	Month	Year

Scottish candidate number

Total marks — 80

SECTION 1 — 20 marks

Attempt ALL questions.

Instructions for the completion of Section 1 are given on *Page two*.

SECTION 2 — 60 marks

Attempt ALL questions.

Necessary Data will be found in the Chemistry Data Booklet for National 5.

Write your answers clearly in the spaces provided in this booklet. Additional space for answers and rough work is provided at the end of this booklet. If you use this space you must clearly identify the question number you are attempting. Any rough work must be written in this booklet. You should score through your rough work when you have written your final copy.

Use **blue** or **black** ink.

Before leaving the examination room you must give this booklet to the Invigilator; if you do not, you may lose all the marks for this paper.

SECTION 1 — 20 marks

The questions for Section 1 are contained in the question paper X713/75/02.

Read these and record your answers on the answer grid on *Page three* opposite.

Use **blue** or **black** ink. Do NOT use gel pens or pencil.

1. The answer to each question is **either** A, B, C or D. Decide what your answer is, then fill in the appropriate bubble (see sample question below).

2. There is **only one correct** answer to each question.

3. Any rough working should be done on the additional space for answers and rough work at the end of this booklet.

Sample Question

To show that the ink in a ball-pen consists of a mixture of dyes, the method of separation would be

    A   fractional distillation

    B   chromatography

    C   fractional crystallisation

    D   filtration.

The correct answer is **B**—chromatography. The answer **B** bubble has been clearly filled in (see below).

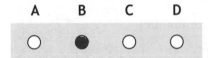

Changing an answer

If you decide to change your answer, cancel your first answer by putting a cross through it (see below) and fill in the answer you want. The answer below has been changed to **D**.

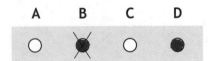

If you then decide to change back to an answer you have already scored out, put a tick (✓) to the **right** of the answer you want, as shown below:

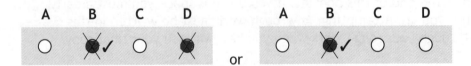

SECTION 1 — Answer Grid

	A	B	C	D
1	○	○	○	○
2	○	○	○	○
3	○	○	○	○
4	○	○	○	○
5	○	○	○	○
6	○	○	○	○
7	○	○	○	○
8	○	○	○	○
9	○	○	○	○
10	○	○	○	○
11	○	○	○	○
12	○	○	○	○
13	○	○	○	○
14	○	○	○	○
15	○	○	○	○
16	○	○	○	○
17	○	○	○	○
18	○	○	○	○
19	○	○	○	○
20	○	○	○	○

SECTION 1 — Answer Grid

[Turn over

[BLANK PAGE]

DO NOT WRITE ON THIS PAGE

[Turn over for next question

DO NOT WRITE ON THIS PAGE

MARKS | DO NOT WRITE IN THIS MARGIN

SECTION 2 — 60 marks

Attempt ALL questions

1. Elements are made up of atoms.

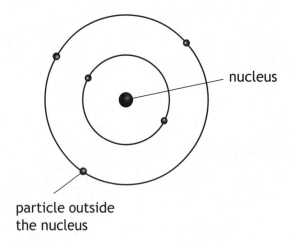

nucleus

particle outside
the nucleus

(a) Complete the tables to show the missing information.

(i)

In the Nucleus		
Particle	*Relative Mass*	*Charge*
proton		+1
neutron	1	

1

(ii)

Outside the Nucleus		
Particle	*Relative Mass*	*Charge*
	almost zero	

1

(b) A sample of nitrogen was found to contain equal amounts of two isotopes. One isotope has mass number 14 and the other has mass number 15.

What is the relative atomic mass of this sample of nitrogen?

1

MARKS | DO NOT WRITE IN THIS MARGIN

1. **(continued)**

(c) Nitrogen can form bonds with other elements.

The diagram shows the shape of a molecule of ammonia (NH_3).

(i) State the name used to describe the shape of a molecule of ammonia.   **1**

(ii) Name the industrial process used to manufacture ammonia.   **1**

[Turn over

MARKS | DO NOT WRITE IN THIS MARGIN

2. The monomer used to produce polystyrene has the following structure.

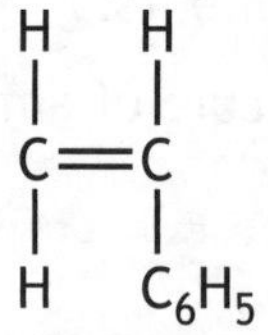

```
    H   H
    |   |
    C = C
    |   |
    H   C₆H₅
      styrene
```

(a) Draw a section of polystyrene, showing three monomer units joined together.

1

(b) When two different monomers polymerise, a copolymer is formed as shown.

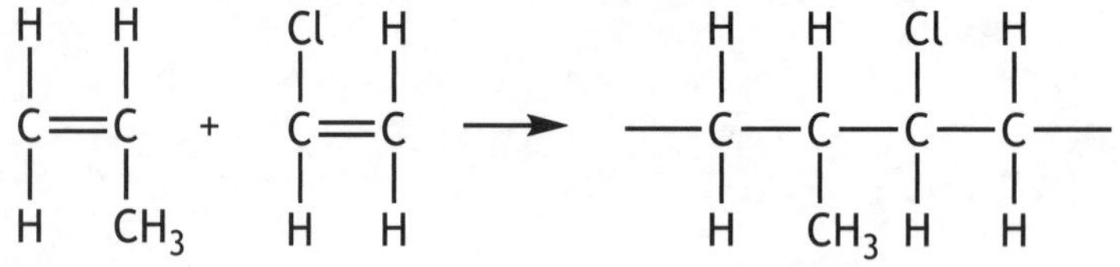

```
  H   H       Cl  H              H   H   Cl  H
  |   |       |   |              |   |   |   |
  C = C   +   C = C   ⟶     — C — C — C — C —
  |   |       |   |              |   |   |   |
  H   CH₃     H   H              H   CH₃ H   H
```

Another copolymer can be made from styrene and acrylonitrile monomers. A section of this copolymer is shown below.

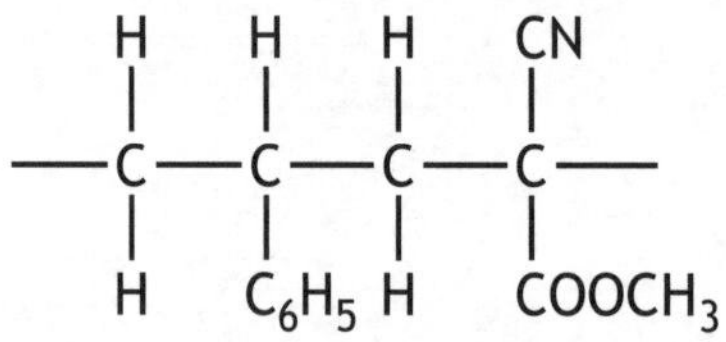

```
      H   H    H   CN
      |   |    |   |
  — C — C — C — C —
      |   |    |   |
      H   C₆H₅ H   COOCH₃
```

Draw the structure of the acrylonitrile monomer.

1

[Turn over for next question

DO NOT WRITE ON THIS PAGE

MARKS | DO NOT WRITE IN THIS MARGIN

3. Hydrogen gas can be produced in the laboratory by adding a metal to dilute acid. Heat energy is also produced in the reaction.

(a) State the term used to describe all chemical reactions that release heat energy.   1

(b) A student measured the volume of hydrogen gas produced when zinc lumps were added to dilute hydrochloric acid.

Time (s)	0	10	20	30	40	50	60	70
Volume of hydrogen (cm^3)	0	12	21	29	34	36	37	37

(i) Calculate the average rate of reaction, in cm^{3}s^{-1}, between 10 and 30 seconds.   2

Show your working clearly.

(ii) Estimate the time taken, in seconds, for the reaction to finish.   1

(iii) The student repeated the experiment using the same mass of zinc.

State the effect on the rate of the reaction if zinc powder was used instead of lumps.   1

MARKS | DO NOT WRITE IN THIS MARGIN

3. (continued)

(c) Another student reacted aluminium with dilute nitric acid.

$$2Al(s) \ + \ 6HNO_3(aq) \longrightarrow 2Al(NO_3)_3(aq) \ + \ 3H_2(g)$$

(i) Circle the formula for the salt in the above equation.   **1**

(ii) 1 mole of hydrogen gas has a volume of 24 litres.

Calculate the volume of hydrogen gas, in litres, produced when 0·01 moles of aluminium react with dilute nitric acid.   **2**

Show your working clearly.

[Turn over

MARKS | DO NOT WRITE IN THIS MARGIN

4. Some rocks contain the mineral with the formula Al_2SiO_5.

This mineral exists in three different forms, andalusite, sillimanite, and kyanite. The form depends on the temperature and pressure.

The diagram shows this relationship.

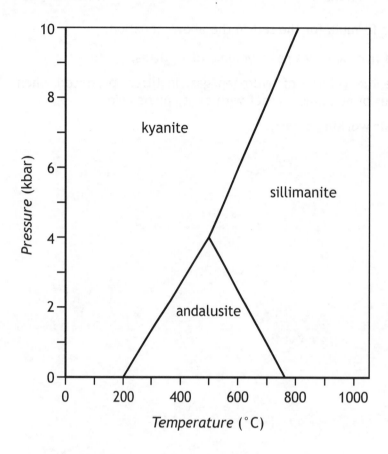

(a) (i) Name the **two** forms which could exist at 400 °C. 1

(ii) Complete the table to show the temperature and pressure at which all three forms would exist. 1

Temperature (°C)	
Pressure (kbar)	

MARKS | DO NOT WRITE IN THIS MARGIN

4.   (continued)

(b) Calculate the percentage mass of silicon in andalusite, Al_2SiO_5.

   Show your working clearly.

3

[Turn over

MARKS | DO NOT WRITE IN THIS MARGIN

5. Read the passage and answer the questions that follow.

Gold—a very useful metal

Gold has been associated with wealth since before the first gold coins were minted in Lydia (modern Turkey) about 550 BC. It does not react with water, air, alkalis and almost all acids. Gold only has one naturally occurring isotope with mass 197.

As an element it has many uses in the modern world. 1 gram of gold can be beaten into a gold film covering one square metre and thin coatings of gold are used as lubricants in aerospace applications. Gold electroplating can be used to coat electrical connectors and printed circuit boards.

Chemists have recently discovered that gold nanoparticles make superb catalysts for many reactions such as the conversion of alcohols into aldehydes and ketones. It can also be used as a catalyst for removing trace carbon monoxide from gases. In this reaction carbon monoxide reacts with oxygen to form carbon dioxide.

Gold nanorods can be grown from a dilute solution of auric acid and are used in the treatment of some forms of cancer.

Adapted from *Education in Chemistry*, Volume 45, November 2008

(a) Suggest a reason why gold was used in the first coins minted.   1

(b) Calculate the number of neutrons present in the naturally occurring isotope of gold.   1

You may wish to use the data booklet to help you.

MARKS | DO NOT WRITE IN THIS MARGIN

5. (continued)

(c) (i) Write an equation, using symbols and formulae, to show the reaction for removing trace carbon monoxide from gases.

There is no need to balance this equation.   1

(ii) State the role of gold in this reaction.   1

(d) Circle the correct words to complete the sentence.   1

Gold nanorods can be grown from a solution which contains

more $\begin{Bmatrix} \text{hydroxide} \\ \text{hydrogen} \end{Bmatrix}$ ions than $\begin{Bmatrix} \text{hydroxide} \\ \text{hydrogen} \end{Bmatrix}$ ions.

[Turn over

MARKS | DO NOT WRITE IN THIS MARGIN

6. (a) A fertiliser for tomato plants contains compounds of phosphorus and potassium.

    (i) Suggest an experimental test, including the result, to show that potassium is present in the fertiliser.

        You may wish to use the data booklet to help you.

        1

    (ii) Ammonium citrate is included in the fertiliser because some phosphorus compounds are more soluble in ammonium citrate solution than they are in water.

        Suggest another reason why ammonium citrate is added to the fertiliser.

        1

(b) In the production of the fertiliser ammonium phosphate, phosphoric acid (H_3PO_4) reacts with ammonium hydroxide as shown.

$$H_3PO_4(aq) \ + \ NH_4OH(aq) \ \longrightarrow \ (NH_4)_3PO_4(aq) \ + \ H_2O(\ell)$$

Balance this equation.

    1

MARKS | DO NOT WRITE IN THIS MARGIN

7. The element strontium was discovered in 1790 in the village of Strontian in Scotland.

   Using your knowledge of chemistry, comment on the chemistry of strontium.

   3

MARKS | DO NOT WRITE IN THIS MARGIN

8. Essential oils can be extracted from plants and used in perfumes and food flavourings.

   (a) Essential oils contain compounds called terpenes.

   A terpene is a chemical made up of a number of isoprene molecules joined together.

   The shortened structural formula of isoprene is $CH_2C(CH_3)CHCH_2$.

   Draw the full structural formula for isoprene.   1

   (b) Essential oils can be extracted from the zest of lemons in the laboratory by steam distillation.

   The process involves heating up water in a boiling tube until it boils. The steam produced then passes over the lemon zest which is separated from the water by glass wool. As the steam passes over the lemon zest it carries the essential oils into a delivery tube. The condensed liquids (essential oils and water) are collected in a test tube placed in a cold water bath.

   Complete the diagram to show the apparatus required to collect the essential oils.   1

   (An additional diagram, if required, can be found on *Page twenty-nine*.)

MARKS | DO NOT WRITE IN THIS MARGIN

8. **(continued)**

(c) Limonene, $C_{10}H_{16}$, is an essential oil which is added to some cleaning products to give them a lemon scent.

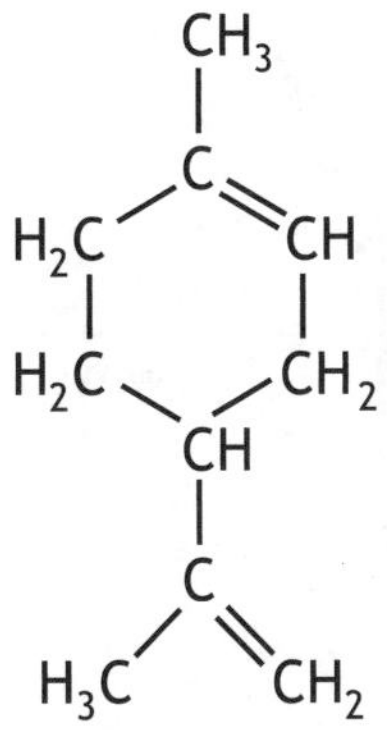

The concentration of limonene present in a cleaning product can be determined by titrating with bromine solution.

(i) Name the type of chemical reaction taking place when limonene reacts with bromine solution.

1

(ii) Write the molecular formula for the product formed when limonene, $C_{10}H_{16}$, reacts completely with bromine solution.

1

[Turn over

MARKS | DO NOT WRITE IN THIS MARGIN

9. Ethanol can be used as an alternative fuel for cars.

   (a) A student considered two methods to confirm the amount of energy released when ethanol burns.

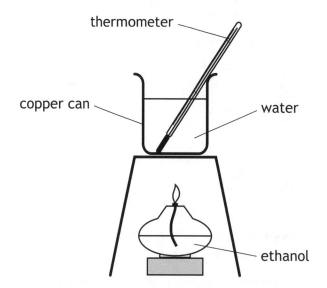

Method **A**	Method **B**
1. Record the initial temperature of the water.	1. Record the initial temperature of the water.
2. Weigh the burner containing the fuel.	2. Weigh the burner containing the fuel.
3. Place the burner under the copper can and then light the burner.	3. Light the burner and then place it under the copper can.
4. Extinguish the flame after 2 minutes.	4. Extinguish the flame after 2 minutes.
5. Record the final temperature and reweigh the burner.	5. Record the final temperature and reweigh the burner.

Explain which method would give a more accurate result.   2

9. **(continued)**

(b) The table gives information about the amount of energy released when 1 mole of some alcohols are burned.

Name of alcohol	Energy released when one mole of alcohol is burned (kJ)
propan-1-ol	2021
propan-2-ol	2005
butan-1-ol	2676
butan-2-ol	2661
pentan-1-ol	3329
pentan-2-ol	3315
hexan-1-ol	3984

(i) Write a statement linking the amount of energy released to the position of the functional group in an alcohol molecule. **1**

(ii) Predict the amount of energy released, in kJ, when 1 mole of hexan-2-ol is burned. **1**

(c) Ethanol can also be used in portable camping stoves.

The chemical reaction in a camping stove releases 23 kJ of energy. If 100 g of water is heated using this stove, calculate the rise in temperature of the water, in °C. **3**

You may wish to use the data booklet to help you.

Show your working clearly.

[Turn over

MARKS | DO NOT WRITE IN THIS MARGIN

10.   A battery is a number of cells joined together.

(a)   The diagram shows a simple battery made from copper and zinc discs separated by paper soaked in potassium nitrate solution.

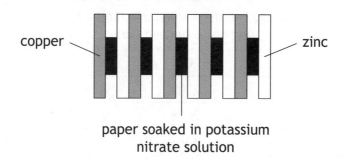

copper

zinc

paper soaked in potassium
nitrate solution

The purpose of the potassium nitrate solution is to complete the circuit.

State the **term** used to describe an ionic compound which is used for this purpose.   1

(b)   A student set up a cell using the same metals as those used in the battery.

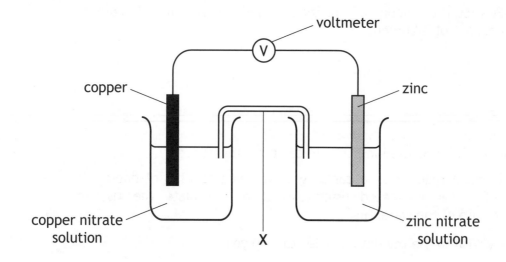

voltmeter

copper

zinc

copper nitrate
solution

X

zinc nitrate
solution

(i)   **On the diagram**, draw an arrow to show the path and direction of electron flow.   1

You may wish to use the data booklet to help you.

(ii) Name the piece of apparatus labelled **X**.   1

MARKS | DO NOT WRITE IN THIS MARGIN

10.   (continued)

(c) Electricity can also be produced in a cell containing non-metals.

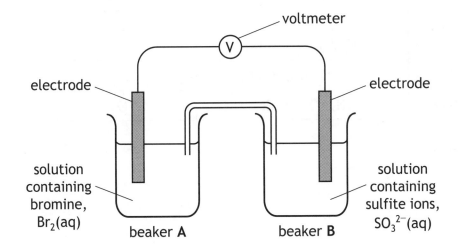

The reactions occurring at each electrode are

Beaker **A** $Br_2(\ell)$ $+$ $2e^-$ $\longrightarrow$ $2Br^-(aq)$

Beaker **B** $SO_3^{2-}(aq) + H_2O(\ell) \longrightarrow SO_4^{2-}(aq) + 2H^+(aq) + 2e^-$

(i) Name the type of chemical reaction taking place in beaker **B**.   1

(ii) Write the redox equation for the overall reaction.   1

(iii) Name a non-metal element which is suitable for use as the electrodes.   1

MARKS | DO NOT WRITE IN THIS MARGIN

11. Ethers are a group of compounds containing carbon, hydrogen and oxygen.

Name of ether	Structural formula	Boiling point (°C)
methoxyethane	$CH_3 - O - CH_2CH_3$	7
ethoxyethane	$CH_3CH_2 - O - CH_2CH_3$	35
X	$CH_3 - O - CH_2CH_2CH_3$	39
propoxybutane	$CH_3CH_2CH_2 - O - CH_2CH_2CH_2CH_3$	117

(a) Name ether X.　　　　1

(b) Suggest a general formula for this homologous series.　　　　1

(c) Methoxyethane is a covalent molecular substance. It has a low boiling point and is a gas at room temperature.

Circle the correct words to complete the sentence.　　　　1

The bonds between the molecules are $\begin{Bmatrix} weak \\ strong \end{Bmatrix}$ and the bonds

within the molecule are $\begin{Bmatrix} weak \\ strong \end{Bmatrix}$.

MARKS | DO NOT WRITE IN THIS MARGIN

11. **(continued)**

(d) Epoxides are a family of cyclic ethers.

The full structural formula for the first member of this family is shown.

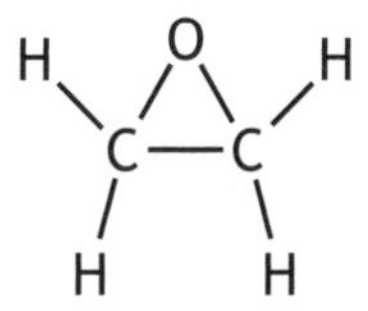

(i) Epoxides can be produced by reacting an alkene with oxygen.

Name the alkene which would be used to produce the epoxide shown.

1

(ii) Epoxides have three atoms in a ring, one of which is oxygen.

Draw a structural formula for the epoxide with the chemical formula C_3H_6O.

1

[Turn over

MARKS | DO NOT WRITE IN THIS MARGIN

12. Betanin is responsible for the red colour in beetroot and can be used as a food colouring.

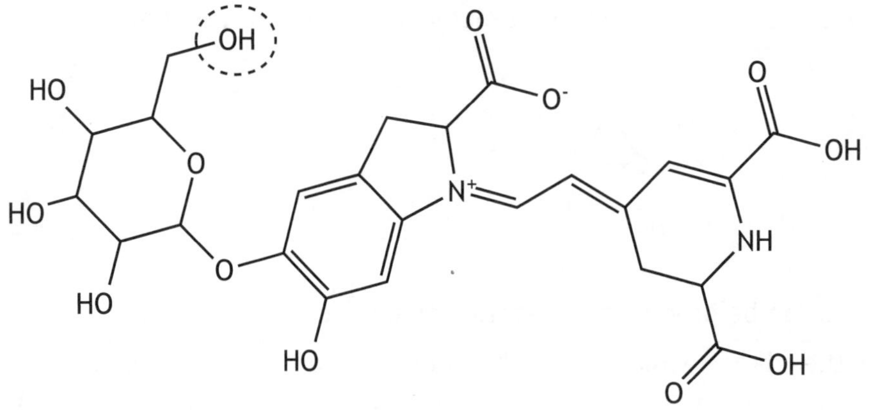

(a) Name the functional group circled in the diagram above.   1

(b) Betanin can be used as an indicator in a neutralisation reaction.

The pH range at which some indicators change colour is shown.

Indicator	pH range of colour change
methyl orange	3·2 to 4·4
litmus	5·0 to 8·0
phenolphthalein	8·2 to 10·0
betanin	9·0 to 10·0

The indicator used in a neutralisation reaction depends on the pH at the end point.

The table below shows the end point of neutralisation reactions using different types of acid and base.

Type of acid	Type of base	pH at the end point
strong	strong	7
strong	weak	below 7
weak	strong	above 7

Betanin can be used to indicate the end point in the reaction between oxalic acid and sodium hydroxide solution.

State the type of acid **and** the type of base used in this reaction.   1

12.   (continued)

(c)   A student carried out a titration experiment to determine the concentration of a sodium hydroxide solution.

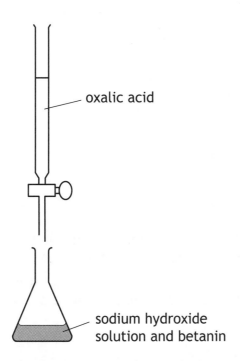

oxalic acid

sodium hydroxide solution and betanin

	Initial burette reading (cm³)	Final burette reading (cm³)	Volume used (cm³)
Rough titre	0·0	15·6	15·6
1st titre	15·6	30·5	14·9
2nd titre	30·5	45·6	15·1

Using the results in the table, calculate the average volume, in cm³, of oxalic acid required to neutralise the sodium hydroxide solution.

1

(d)   Oxalic acid is found naturally in rhubarb. A piece of rhubarb was found to contain 1·8 g of oxalic acid.

Calculate the number of moles of oxalic acid contained in the piece of rhubarb.

(Formula mass of oxalic acid = 90)

1

[Turn over for next question

MARKS | DO NOT WRITE IN THIS MARGIN

13. Carbonated water, also known as sparkling water, is water into which carbon dioxide gas has been dissolved. This process is called carbonating.

A group of students are given two brands of carbonated water and asked to determine which brand contains more dissolved carbon dioxide.

Using your knowledge of chemistry, describe how the students could determine which brand of carbonated water contains more dissolved carbon dioxide.

3

[END OF QUESTION PAPER]

MARKS | DO NOT WRITE IN THIS MARGIN

ADDITIONAL SPACE FOR ANSWERS

Additional diagram for Question 8 (b)

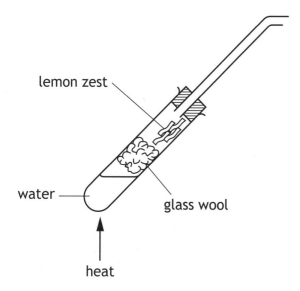

MARKS | DO NOT WRITE IN THIS MARGIN

ADDITIONAL SPACE FOR ANSWERS AND ROUGH WORK

MARKS | DO NOT WRITE IN THIS MARGIN

ADDITIONAL SPACE FOR ANSWERS AND ROUGH WORK

[BLANK PAGE]

DO NOT WRITE ON THIS PAGE

NATIONAL 5

2017

MONDAY, 8 MAY

1:00 PM — 3:00 PM

Instructions for the completion of Section 1 are given on *Page two* of your question and answer booklet X713/75/01.

Record your answers on the answer grid on *Page three* of your question and answer booklet.

You may refer to the Chemistry Data Booklet for National 5.

Before leaving the examination room you must give your question and answer booklet to the Invigilator; if you do not, you may lose all the marks for this paper.

SECTION 1

1. In a reaction, the mass lost in 30 seconds was 2 g.

   What is the average rate of reaction, in $g\,s^{-1}$, over this time?

   A $\dfrac{1}{30}$

   B $\dfrac{30}{2}$

   C $\dfrac{1}{2}$

   D $\dfrac{2}{30}$

2. An atom has 21 protons, 21 electrons and 24 neutrons.

   The atom has

   A atomic number 24 and mass number 42

   B atomic number 45 and mass number 21

   C atomic number 21 and mass number 45

   D atomic number 24 and mass number 45.

3. What is the charge on the zinc ion in zinc dichromate, $ZnCr_2O_7$?

   You may wish to use the data booklet to help you.

   A 2+

   B 2−

   C 1+

   D 1−

4. The table contains information about magnesium and magnesium chloride.

	Melting Point ($^{\circ}C$)	Density ($g\,cm^{-3}$)
Magnesium	650	1·74
Magnesium chloride	714	2·32

   When molten magnesium chloride is electrolysed at 730 $^{\circ}$C the magnesium appears as a

   A solid on the surface of the molten magnesium chloride

   B solid at the bottom of the molten magnesium chloride

   C liquid at the bottom of the molten magnesium chloride

   D liquid on the surface of the molten magnesium chloride.

5. Which of the following compounds is a base?

     A   Sodium carbonate

     B   Sodium chloride

     C   Sodium nitrate

     D   Sodium sulfate

6. $AgNO_3(aq) \; + \; KCl(aq) \; \longrightarrow \; AgCl(s) \; + \; KNO_3(aq)$

   Which of the following are the spectator ions in this reaction?

     A Ag^+ and Cl^-

     B K^+ and NO_3^-

     C Ag^+ and NO_3^-

     D K^+ and Cl^-

7. $x\,H_2O_2 \; \longrightarrow \; y\,H_2O \; + \; z\,O_2$

   This equation will be balanced when

     A $x = 1$, $y = 2$ and $z = 2$

     B $x = 1$, $y = 1$ and $z = 2$

     C $x = 2$, $y = 2$ and $z = 1$

     D $x = 2$, $y = 2$ and $z = 2$.

8. 0·25 moles of a gas has a mass of 7 g.

   Which of the following could be the molecular formula for the gas?

     A C_2H_6

     B C_2H_4

     C C_3H_8

     D C_3H_6

9. Which of the following solutions contains the **least** number of moles of solute?

     A $100\,cm^3$ of $0·4\,mol\,l^{-1}$ solution

     B $200\,cm^3$ of $0·3\,mol\,l^{-1}$ solution

     C $300\,cm^3$ of $1·0\,mol\,l^{-1}$ solution

     D $400\,cm^3$ of $0·5\,mol\,l^{-1}$ solution

[Turn over

10. Which of the following could be the molecular formula for an alkane?

A C_7H_{16}

B C_7H_{14}

C C_7H_{12}

D C_7H_{10}

11. A student added bromine solution to compound X and compound Y.

Compound X   Compound Y

Which line in the table is correct?

	Decolourises bromine solution immediately	
	Compound X	Compound Y
A	no	no
B	no	yes
C	yes	yes
D	yes	no

12. A compound burns in air. The only products of the reaction are carbon dioxide, sulfur dioxide and water.

The compound **must** contain

A carbon and sulfur only

B carbon and hydrogen only

C carbon, hydrogen and sulfur

D carbon, hydrogen, sulfur and oxygen.

13. Vinegar is a solution of

A ethanol

B methanol

C ethanoic acid

D methanoic acid.

14. A reaction is exothermic if

   A   energy is absorbed from the surroundings

   B   energy is released to the surroundings

   C   energy is required to start the reaction

   D   there is no energy change.

15. Which of the following diagrams could be used to represent the structure of copper?

   A

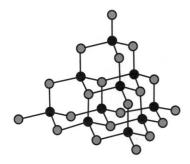

   B

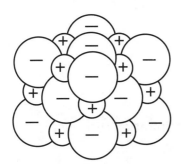

   C

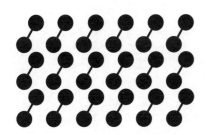

   D

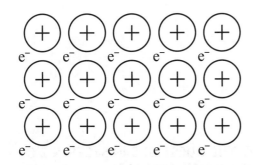

[Turn over

16. Which of the following metals is found uncombined in the Earth's crust?
    You may wish to use the data booklet to help you.

    A   Tin
    B   Magnesium
    C   Gold
    D   Sodium

17. Which of the following is **not** an essential element for healthy plant growth?

    A   Oxygen
    B   Nitrogen
    C   Potassium
    D   Phosphorus

18. The Haber process is the industrial process for the manufacture of

    A   nitric acid
    B   ammonia
    C   alkenes
    D   esters.

19. Which of the following salts can be prepared by a precipitation reaction?
    You may wish to use the data booklet to help you.

    A   Barium sulfate
    B   Lithium nitrate
    C   Calcium chloride
    D   Ammonium phosphate

20. A solution of accurately known concentration is more commonly known as a

    A   correct solution
    B   precise solution
    C   standard solution
    D   prepared solution.

[END OF SECTION 1. NOW ATTEMPT THE QUESTIONS IN SECTION 2 OF
YOUR QUESTION AND ANSWER BOOKLET]

[BLANK PAGE]

DO NOT WRITE ON THIS PAGE

[BLANK PAGE]

DO NOT WRITE ON THIS PAGE

FOR OFFICIAL USE

N5

National
Qualifications
2017

Mark

X713/75/01

**Chemistry
Section 1—Answer Grid
And Section 2**

MONDAY, 8 MAY

1:00 PM – 3:00 PM

Fill in these boxes and read what is printed below.

Full name of centre

Town

Forename(s)

Surname

Number of seat

Date of birth
Day Month Year

Scottish candidate number

Total marks — 80

SECTION 1 — 20 marks

Attempt ALL questions.

Instructions for the completion of Section 1 are given on *Page two*.

SECTION 2 — 60 marks

Attempt ALL questions.

You may refer to the Chemistry Data Booklet for National 5.

Write your answers clearly in the spaces provided in this booklet. Additional space for answers and rough work is provided at the end of this booklet. If you use this space you must clearly identify the question number you are attempting. Any rough work must be written in this booklet. You should score through your rough work when you have written your final copy.

Use **blue** or **black** ink.

Before leaving the examination room you must give this booklet to the Invigilator; if you do not, you may lose all the marks for this paper.

SECTION 1— 20 marks

The questions for Section 1 are contained in the question paper X713/75/02.

Read these and record your answers on the answer grid on *Page three* opposite.

Use **blue** or **black** ink. Do NOT use gel pens or pencil.

1. The answer to each question is **either** A, B, C or D. Decide what your answer is, then fill in the appropriate bubble (see sample question below).

2. There is **only one correct** answer to each question.

3. Any rough working should be done on the additional space for answers and rough work at the end of this booklet.

Sample Question

To show that the ink in a ball-pen consists of a mixture of dyes, the method of separation would be

    A   fractional distillation

    B   chromatography

    C   fractional crystallisation

    D   filtration.

The correct answer is **B** — chromatography. The answer **B** bubble has been clearly filled in (see below).

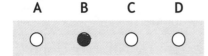

Changing an answer

If you decide to change your answer, cancel your first answer by putting a cross through it (see below) and fill in the answer you want. The answer below has been changed to **D**.

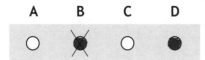

If you then decide to change back to an answer you have already scored out, put a tick (✓) to the **right** of the answer you want, as shown below:

SECTION 1 — Answer Grid

[Turn over

	A	B	C	D
1	○	○	○	○
2	○	○	○	○
3	○	○	○	○
4	○	○	○	○
5	○	○	○	○
6	○	○	○	○
7	○	○	○	○
8	○	○	○	○
9	○	○	○	○
10	○	○	○	○
11	○	○	○	○
12	○	○	○	○
13	○	○	○	○
14	○	○	○	○
15	○	○	○	○
16	○	○	○	○
17	○	○	○	○
18	○	○	○	○
19	○	○	○	○
20	○	○	○	○

[BLANK PAGE]

DO NOT WRITE ON THIS PAGE

MARKS | DO NOT WRITE IN THIS MARGIN

SECTION 2 — 60 marks

Attempt ALL questions

1. A sample of argon contains three types of atom.

$$^{36}_{18}Ar \qquad ^{38}_{18}Ar \qquad ^{40}_{18}Ar$$

(a) State the term used to describe these different types of argon atom.

1

(b) Explain why the mass number of each type of atom is different.

1

(c) This sample of argon has an average atomic mass of 36·2.

State the mass number of the most common type of atom in the sample of argon.

1

[Turn over

MARKS | DO NOT WRITE IN THIS MARGIN

2. Read the passage below and attempt the questions that follow.

Hydrogen Storage

The portable storage of hydrogen (H_2) is key to the development of hydrogen fuel cell cars. While many chemists focus their attention on the use of metal alloys and hydrides for storing hydrogen, others have investigated the potential use of carbon nanotubes.

A carbon nanotube is a tiny rolled up sheet of graphite. A research team has designed a pillared structure made up of vertical columns of carbon nanotubes which stabilise parallel graphene sheets. Graphene sheets are layers of carbon which are one atom thick.

Lithium atoms are added to the pillared structure to increase the hydrogen storage capacity. Researchers claim that one litre of the structure can store 41 g of hydrogen, which comes close to the US Department of Energy's target of 45 g.

Adapted from *InfoChem Magazine* (RSC), Nov 2008

(a) Name the term used to describe a tiny rolled up sheet of graphite.   **1**

(b) Name the metal added to the pillared structure to increase the hydrogen storage capacity.   **1**

(c) Calculate the number of moles of hydrogen that, researchers claim, can be stored by one litre of this structure.   **2**

   Show your working clearly.

[Turn over for next question

DO NOT WRITE ON THIS PAGE

MARKS | DO NOT WRITE IN THIS MARGIN

3. Chlorine can form covalent and ionic bonds.

   (a) Chlorine gas is made up of diatomic molecules.

   Draw a diagram, showing all outer electrons, to represent a molecule of chlorine, Cl_2.

   1

   (b) Chloromethane is a covalent gas with a faint sweet odour.

   The structure of a chloromethane molecule is shown.

   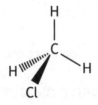

   State the name used to describe the shape of a molecule of chloromethane.

   1

MARKS | DO NOT WRITE IN THIS MARGIN

3. (continued)

(c) When chlorine reacts with sodium the ionic compound sodium chloride is formed.

A chloride ion has a stable electron arrangement.

Describe how a chlorine atom achieves this stable electron arrangement.

1

(d) Covalent and ionic compounds have different physical properties.

Complete the table by circling the words which correctly describe the properties of the two compounds.

2

Compound	Melting point	Conductor of electricity
chloromethane gas	high/low	yes/no
solid sodium chloride	high/low	yes/no

[Turn over

MARKS | DO NOT WRITE IN THIS MARGIN

4. Iron is produced from iron ore in a blast furnace.

    (a) Iron ore, limestone and carbon are added at the top of the blast furnace. Hot air is blown in near the bottom of the furnace and, through a series of chemical reactions, iron is produced. Waste gases are released near the top of the furnace. A layer of impurities is also produced which floats on top of the iron. The iron and impurities both flow off separately at the bottom of the furnace.

        (i) Use this information to complete the diagram.   2

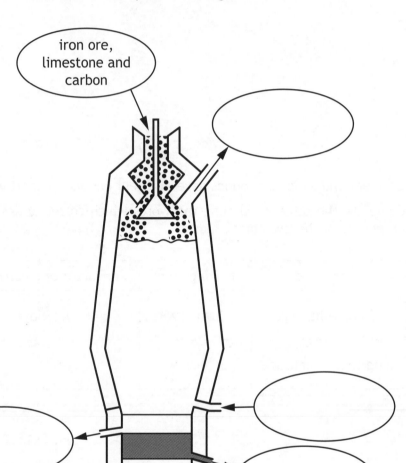

MARKS | DO NOT WRITE IN THIS MARGIN

4. (a) (continued)

(ii) Explain why the temperature at the bottom of the blast furnace should not drop below 1538 °C.   **1**

You may wish to use the data booklet to help you.

(b) Rusting occurs when iron is exposed to air and water.

During rusting, iron initially loses two electrons to form iron(II) ions. These ions are further oxidised to form iron(III) ions.

Write an ion-electron equation to show iron(II) ions forming iron(III) ions.   **1**

You may wish to use the data booklet to help you.

[Turn over

MARKS | DO NOT WRITE IN THIS MARGIN

5. Phosphorus-32 is a radioisotope used in the detection of cancerous tumours.

(a) The graph shows how the percentage of phosphorus-32 in a sample changes over a period of time.

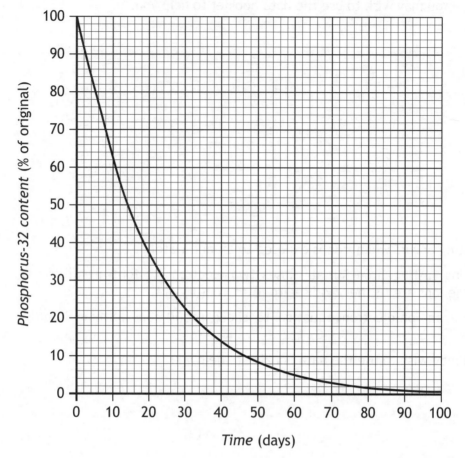

Time (days)

(i) Using the graph, calculate the half-life, in days, of phosphorus-32.   **1**

(ii) Using your answer to part (a) (i), calculate the time, in days, it would take for the mass of a 20 g sample of the radioisotope to decrease to 2·5 g.   **2**

(b) Phosphorus-32 decays by emitting radiation.

During this decay the atomic number increases by 1.

Name the type of radiation emitted when phosphorus-32 decays.   **1**

MARKS | DO NOT WRITE IN THIS MARGIN

6. A student wanted to investigate whether copper could be used as a catalyst for the reaction between zinc and sulfuric acid.

$$Zn(s) \ + \ H_2SO_4(aq) \longrightarrow ZnSO_4(aq) \ + \ H_2(g)$$

Using your knowledge of chemistry, suggest how the student could investigate this.

3

[Turn over

MARKS | DO NOT WRITE IN THIS MARGIN

7. Carboxylic acids can be used in household cleaning products.

   (a) Name the functional group found in all carboxylic acids. **1**

   (b) Carboxylic acids have a range of physical and chemical properties. Melting point is an example of a physical property.

   The table gives information about propanoic acid and butanoic acid.

Carboxylic acid	Melting point (°C)
propanoic acid	−21
butanoic acid	−5

   (i) Draw a structural formula for butanoic acid. **1**

   (ii) Explain why butanoic acid has a higher melting point than propanoic acid. **2**

MARKS | DO NOT WRITE IN THIS MARGIN

8.   A teacher demonstrated the following experiment.

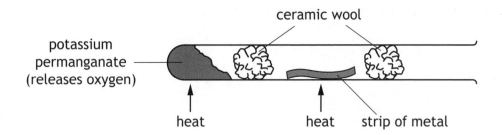

The results are shown in the table.

Metal	Observation
zinc	glowed brightly
copper	dull red glow
silver	no reaction

(a)   (i)   Describe what would be observed if the experiment was repeated using magnesium.   1

(ii) The teacher repeated the experiment using copper powder.

State the effect this would have on the rate of the reaction between copper and oxygen.   1

(b) Magnesium also reacts with steam to produce magnesium oxide and hydrogen gas.

$$Mg(s) \quad + \quad H_2O(g) \quad \longrightarrow \quad MgO(s) \quad + \quad H_2(g)$$

Identify the substance which is being oxidised.   1

[Turn over

Page fifteen

MARKS | DO NOT WRITE IN THIS MARGIN

9. The alkanes are a homologous series of saturated hydrocarbons.

   (a) State what is meant by the term homologous series.   1

   (b) The structural formula of two alkanes is shown.

2-methylpentane   2,3-dimethylbutane

   State the term used to describe a pair of alkanes such as 2-methylpentane and 2,3-dimethylbutane.   1

[Turn over

MARKS | DO NOT WRITE IN THIS MARGIN

9. (continued)

(c) The alkanes present in a mixture were separated using a technique known as HPLC. The mixture was vaporised and then passed through a special column. Different alkanes take different amounts of time to pass through the column.

The results are shown.

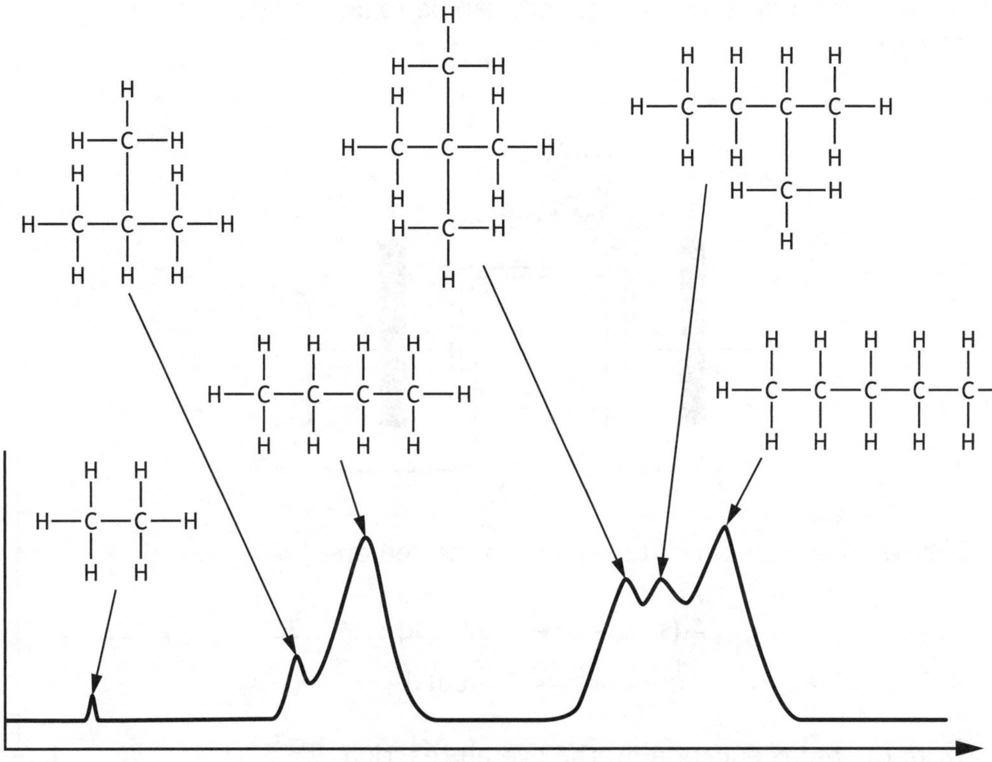

Time taken to pass through the column

(i) Write a general statement linking the structure of the alkane to the length of time taken to pass through the column. **1**

(ii) Propane was added to the mixture and the HPLC technique was repeated.

Draw an arrow on the graph to show the expected time taken for propane to pass through the column. **1**

(An additional diagram, if required, can be found on *Page twenty-seven*.)

10. A student set up an electrochemical cell using aluminium and copper electrodes as well as aluminium sulfate solution and copper(II) sulfate solution.

(a) (i) Complete the labels on the diagram to show the electrochemical cell which would give the direction of electron flow indicated.   1

You may wish to use the data booklet to help you.

(An additional diagram, if required, can be found on *Page twenty-seven.*)

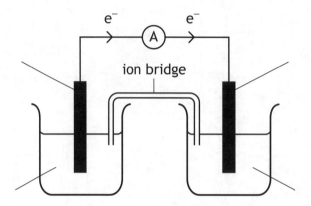

(ii) The two reactions which take place in the cell are

$$Al(s) \longrightarrow Al^{3+}(aq) + 3e^-$$

$$Cu^{2+}(aq) + 2e^- \longrightarrow Cu(s)$$

Write the redox equation for the overall reaction.   1

(b) Calculate the percentage by mass of aluminium in aluminium sulfate, $Al_2(SO_4)_3$.   3

Show your working clearly.

[Turn over

11. Sulfur dioxide is an important industrial chemical.

Sulfur dioxide dissolves in water to produce sulfurous acid.

$$SO_2(g) \quad + \quad H_2O(\ell) \quad \longrightarrow \quad H_2SO_3(aq)$$

(a) Explain the change in the pH of the solution as sulfur dioxide dissolves.   **2**

(b) The graph shows the solubility of sulfur dioxide at different temperatures.

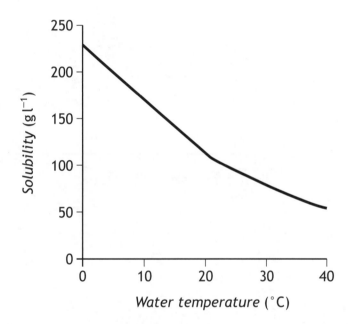

Describe the general trend in solubility as the temperature of the water increases.   **1**

MARKS | DO NOT WRITE IN THIS MARGIN

12. Geraniol is an essential oil known to have anti-inflammatory properties. A structure for the geraniol molecule is shown.

$$H_3C \diagdown \quad \diagup CH_2OH$$
$$C=C$$
$$H_2C \diagup \quad \diagdown H$$
$$|$$
$$H_2C \diagdown \quad \diagup H$$
$$C$$
$$||$$
$$C$$
$$H_3C \diagup \quad \diagdown CH_3$$

(a) Circle a functional group found in the geraniol molecule.   1

(An additional diagram, if required, can be found on *Page twenty-eight*.)

[Turn over

MARKS | DO NOT WRITE IN THIS MARGIN

12. **(continued)**

(b) One of the compounds used to flavour foods is geranyl propanoate.

Name the family to which geranyl propanoate belongs.   **1**

(c) A student prepared a sample of geranyl propanoate from geraniol and propanoic acid.

geraniol + propanoic acid $\longrightarrow$ geranyl propanoate + water

$C_{10}H_{18}O$ + $C_3H_6O_2$ $\longrightarrow$ $C_{13}H_{22}O_2$ + H_2O

15·4 g of geraniol was reacted with excess propanoic acid.

Calculate the mass, in grams, of geranyl propanoate which would be produced.   **3**

Show your working clearly.

[Turn over

MARKS | DO NOT WRITE IN THIS MARGIN

13. The alkynes are a family of hydrocarbons which contain a carbon to carbon triple bond. Three members of this family are shown.

propyne   but-1-yne   pent-1-yne

(a) Suggest a general formula for the alkyne family.   1

(b) Ethyne can undergo polymerisation to form poly(ethyne).

$$H-C\equiv C-H \longrightarrow$$

(i) Draw the repeating unit in the polymer poly(ethyne).   1

(ii) Name the type of polymerisation taking place when ethyne is converted to poly(ethyne).   1

[Turn over

MARKS | DO NOT WRITE IN THIS MARGIN

13.   (continued)

(c)   Alkynes can be prepared by reacting a dibromoalkane with potassium hydroxide solution.

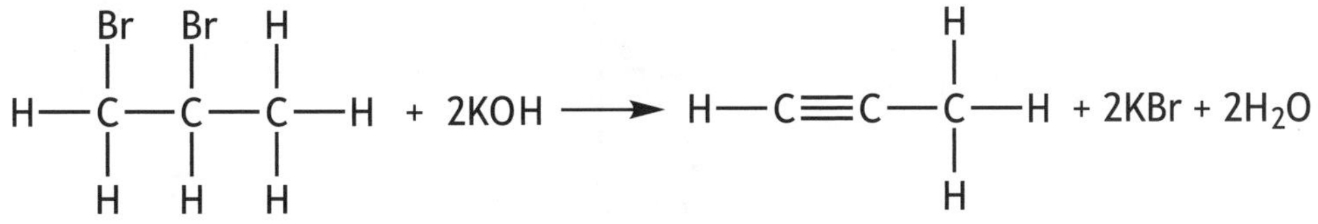

1,2-dibromopropane   propyne

(i)   Draw the **full** structural formula for the alkyne formed when 2,3-dibromobutane reacts with potassium hydroxide.   **1**

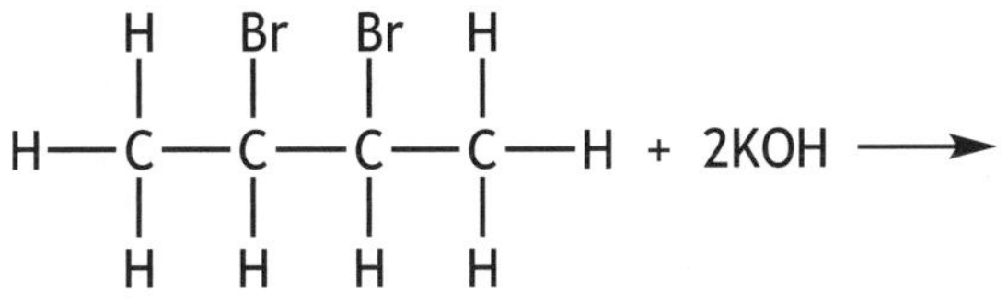

2,3-dibromobutane

(ii)   The structure for 2,4-dibromopentane is shown below.

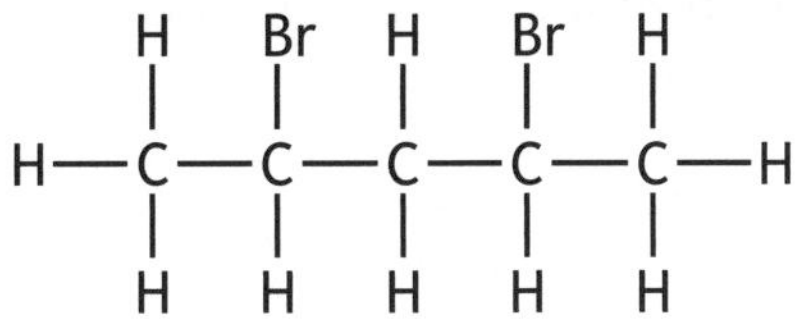

2,4-dibromopentane

Suggest a reason why 2,4-dibromopentane does **not** form an alkyne when it is added to potassium hydroxide solution.   **1**

[Turn over

MARKS | DO NOT WRITE IN THIS MARGIN

14. (a) A group of students carried out an experiment to measure the energy produced when 5 g samples of different alcohols were burned.

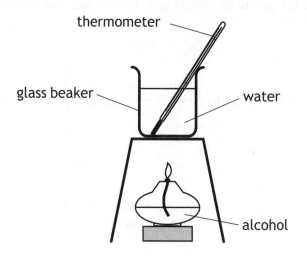

The results are shown.

Alcohol	Energy released (kJ)
propan-1-ol	158
butan-1-ol	170
pentan-1-ol	179
hexan-1-ol	185

(i) Draw a structural formula for hexan-1-ol.   1

(ii) Predict the energy released, in kJ, if the same mass of heptan-1-ol was burned.   1

[Turn over

MARKS | DO NOT WRITE IN THIS MARGIN

14. (continued)

(b) The energy released when an alcohol burns can be used to heat liquids other than water.

The data below was collected when the energy released, by burning an alcohol, was used to heat a sodium chloride solution.

Energy released when the alcohol was burned (kJ)	13·3
Initial temperature of sodium chloride solution (°C)	15
Final temperature of sodium chloride solution (°C)	49
Mass of sodium chloride solution heated (g)	100

Calculate the specific heat capacity, in $kJ\,kg^{-1}\,°C^{-1}$, of the sodium chloride solution.

You may wish to use the data booklet to help you.

Show your working clearly.

3

[Turn over for next question

MARKS | DO NOT WRITE IN THIS MARGIN

15. A student was given two solutions of sodium carbonate, one solution with a concentration of $0.1\,mol\,l^{-1}$ and the other with a concentration of $0.2\,mol\,l^{-1}$.

Using your knowledge of chemistry, suggest how the student could distinguish between the solutions.

3

[END OF QUESTION PAPER]

ADDITIONAL SPACE FOR ANSWERS

Additional diagram for Question 9 (c) (ii)

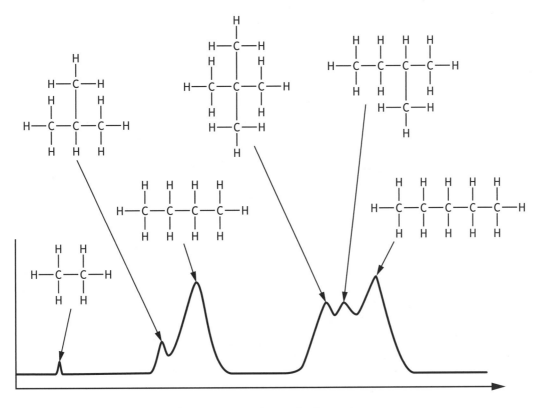

Time taken to pass through the column

Additional diagram for Question 10 (a) (i)

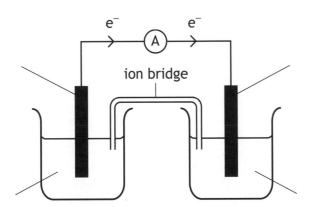

MARKS | DO NOT WRITE IN THIS MARGIN

ADDITIONAL SPACE FOR ANSWERS

Additional diagram for Question 12 (a)

ADDITIONAL SPACE FOR ANSWERS AND ROUGH WORK

[BLANK PAGE]

DO NOT WRITE ON THIS PAGE

NATIONAL 5

2017 Specimen
Question Paper

S813/75/02

Chemistry
Section 1—Questions

Date — Not applicable

Duration — 2 hours 30 minutes

Instructions for completion of Section 1 are given on *Page two* of your question and answer booklet S813/75/01.

Record your answers on the answer grid on *Page three* of your question and answer booklet.

You may refer to the Chemistry Data Booklet for National 5.

Before leaving the examination room you must give your question and answer booklet to the Invigilator; if you do not, you may lose all the marks for this paper.

SECTION 1 — 25 marks

Attempt ALL questions

1. Which of the following elements usually exists as diatomic molecules?

   A   Helium

   B   Nitrogen

   C   Silicon

   D   Sulfur

2. Which line in the table correctly describes a proton?

	Mass (atomic mass units)	Charge
A	negligible	+1
B	negligible	−1
C	1	+1
D	1	0

3. Ionic compounds conduct electricity when molten because they have

   A   ions that are free to move

   B   delocalised electrons

   C   metal atoms

   D   a lattice structure.

4. A molecule of phosphine is shown below.

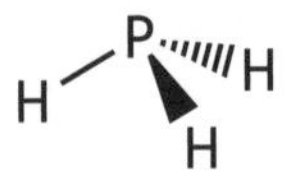

   The shape of a molecule of phosphine is

   A   linear

   B   angular

   C   tetrahedral

   D   trigonal pyramidal.

5. The table gives information about some particles.

Identify the particle which is a negative ion.

Particle	Number of		
	protons	neutrons	electrons
A	9	10	10
B	11	12	11
C	15	16	15
D	19	20	18

6. The table shows the colours of some ionic compounds in solution.

Compound	Colour
copper nitrate	blue
copper chromate	green
strontium nitrate	colourless
strontium chromate	yellow

The colour of the chromate ion is

A blue

B green

C colourless

D yellow.

7. Which of the following statements correctly describes the concentrations of $H^+(aq)$ and $OH^-(aq)$ ions in pure water?

A The concentrations of $H^+(aq)$ and $OH^-(aq)$ ions are equal.

B The concentrations of $H^+(aq)$ and $OH^-(aq)$ ions are zero.

C The concentration of $H^+(aq)$ ions is greater than the concentration of $OH^-(aq)$ ions.

D The concentration of $H^+(aq)$ ions is less than the concentration of $OH^-(aq)$ ions.

[Turn over

8.

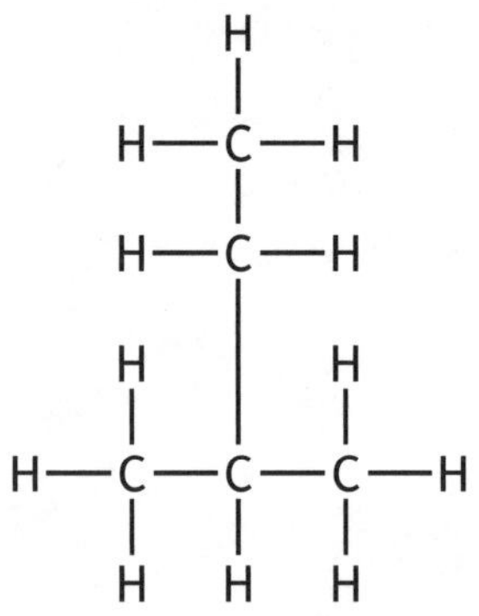

The name of the above compound is

A   2-ethylpropane

B   1,1-dimethylpropane

C   2-methylbutane

D   3-methylbutane.

9. Which of the following could be the molecular formula for a cycloalkane?

A C_6H_8

B C_6H_{10}

C C_6H_{12}

D C_6H_{14}

10. In which of the following types of reaction is oxygen a reactant?

A   Combustion

B   Neutralisation

C   Polymerisation

D   Precipitation

11. Molecules in which four different atoms are attached to a carbon atom are said to be chiral.

Which of the following molecules is chiral?

A

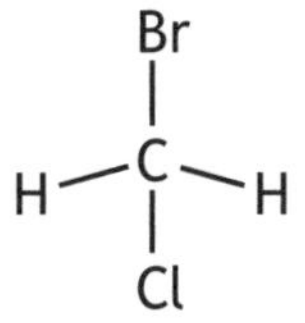

$$H-\overset{\overset{\displaystyle Br}{|}}{\underset{\underset{\displaystyle Cl}{|}}{C}}-H$$

B

$$H-\overset{\overset{\displaystyle Br}{|}}{\underset{\underset{\displaystyle H}{|}}{C}}-H$$

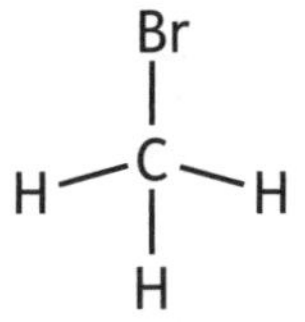

C

$$Cl-\overset{\overset{\displaystyle I}{|}}{\underset{\underset{\displaystyle H}{|}}{C}}-H$$

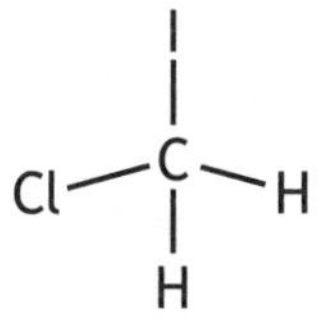

D

$$H-\overset{\overset{\displaystyle I}{|}}{\underset{\underset{\displaystyle F}{|}}{C}}-Br$$

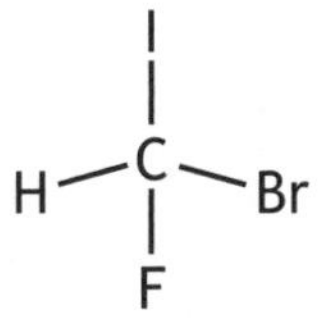

12. Three members of the cycloalkene family are

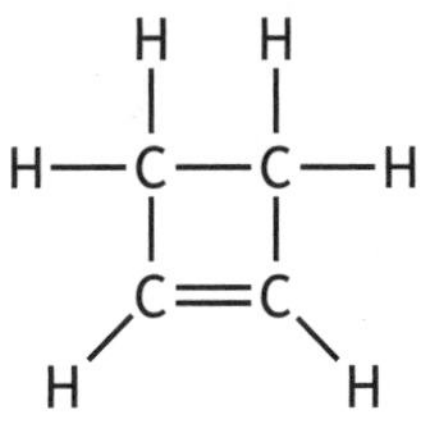

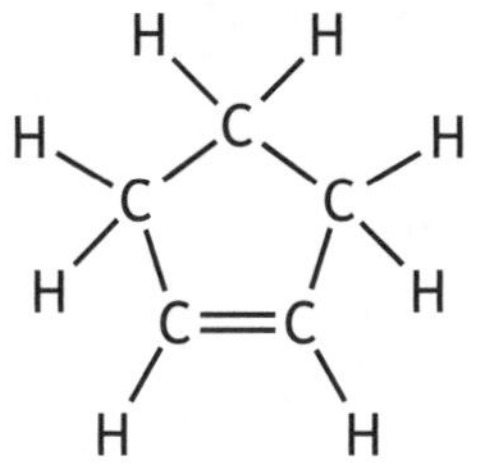

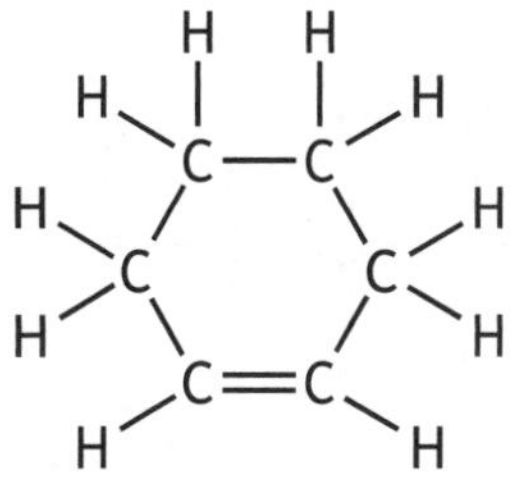

The general formula for the cycloalkene family is

A C_nH_{2n-2}

B C_nH_{2n-4}

C C_nH_{2n}

D C_nH_{2n+2}

[Turn over

13. Which of the following molecules is an isomer of hept-2-ene?

A

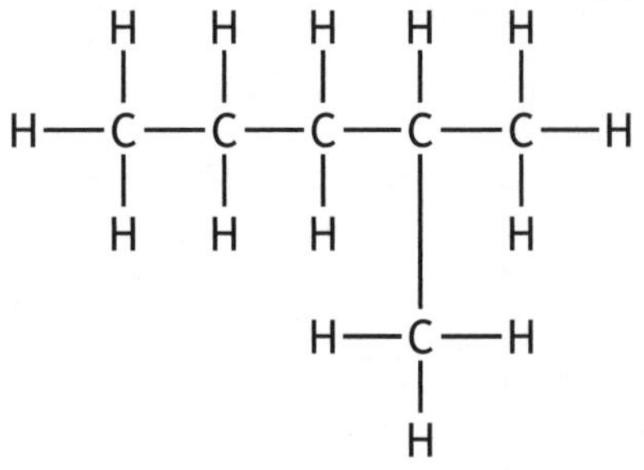

B

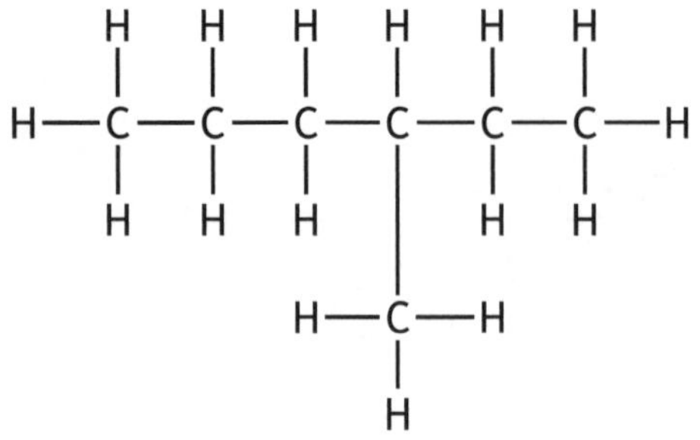

C

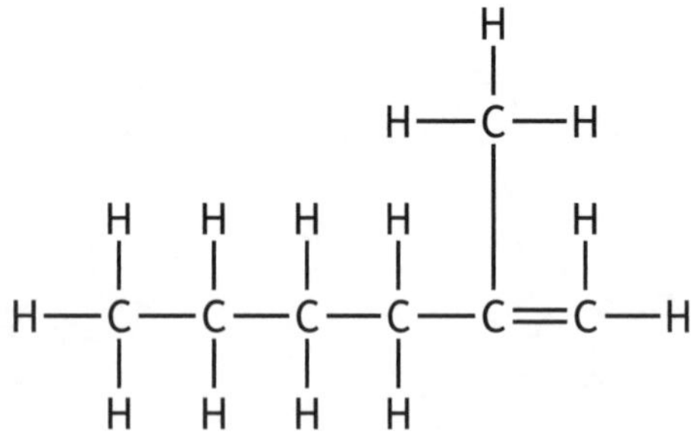

D

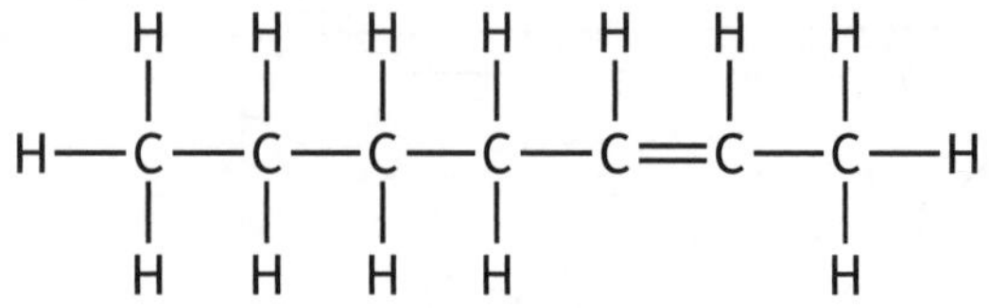

14. A student tested some compounds. The results are given in the table.

Compound	pH of aqueous solution	Effect on bromine solution
H—C(H)(H)—C(H)(H)—C(=O)(OH)	4	no effect
H—C=C(H)—C(=O)(OH) with H,H	4	decolourised
H—C(H)(H)—C(H)(H)—C(H)(H)—OH	7	no effect
H—C=C(H)—C(H)(H)—OH with H	7	decolourised

Which line in the table below shows the correct results for the following compound?

H—C(H)(H)—C(H)=C(H)—C(H)(H)—C(H)(H)—OH

	pH of aqueous solution	Effect on bromine solution
A	4	decolourised
B	7	decolourised
C	4	no effect
D	7	no effect

[Turn over

15. Which of the following diagrams could be used to represent the structure of a metal?

A

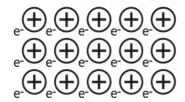

B

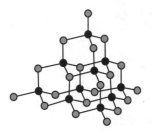

C

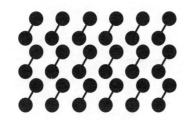

D

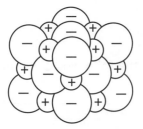

16. Which of the following substances does **not** produce water when it reacts with dilute acid?

A   Sodium hydroxide

B   Magnesium

C   Copper oxide

D   Ammonia solution

17. Which of the following metals can be extracted from its oxide by heat alone?

A   Aluminium

B   Zinc

C   Gold

D   Iron

18.

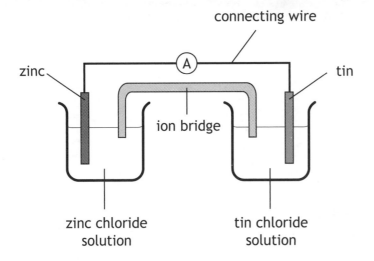

In the cell shown above, electrons flow through

A   the solution from tin to zinc

B   the solution from zinc to tin

C   the connecting wire from tin to zinc

D   the connecting wire from zinc to tin.

19.   Four cells were made by joining silver to copper, iron, tin and zinc.

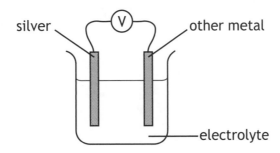

The voltages for the four cells are shown in the table.

Which cell contained silver joined to copper?

You may wish to use the data booklet to help you.

Cell	Voltage (V)
A	1·6
B	1·2
C	0·9
D	0·5

[Turn over

20. The ion-electron equation for the oxidation and reduction steps in the reaction between magnesium and silver(I) ions are:

$$Mg \rightarrow Mg^{2+} + 2e^-$$

$$Ag^+ + e^- \rightarrow Ag$$

The overall redox equation is

A $Mg + 2Ag^+ \rightarrow Mg^{2+} + 2Ag$

B $Mg + Ag^+ \rightarrow Mg^{2+} + Ag$

C $Mg + Ag^+ + e^- \rightarrow Mg^{2+} + Ag + 2e^-$

D $Mg + 2Ag \rightarrow Mg^{2+} + 2Ag^+$.

21. The structure below shows a section of an addition polymer.

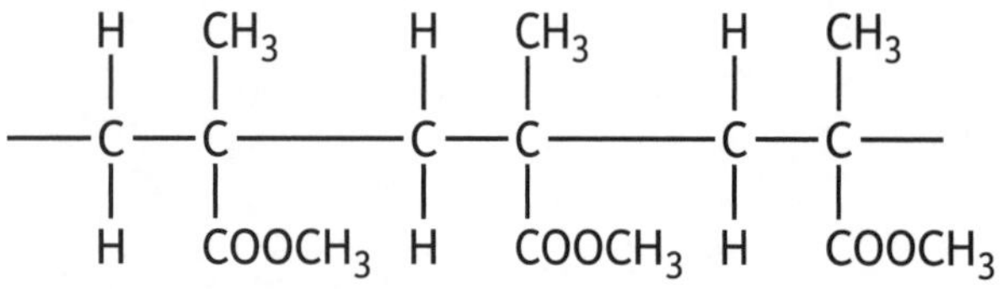

Which of the following molecules is used to make this polymer?

A

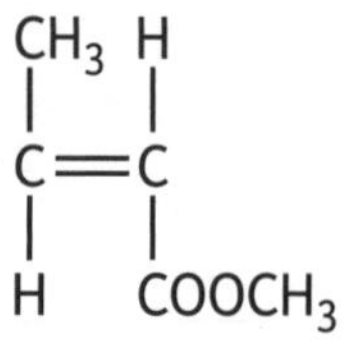

B

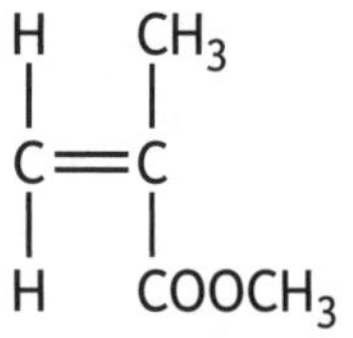

C

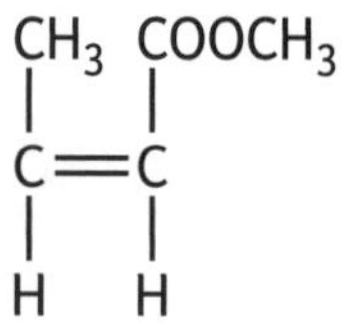

D

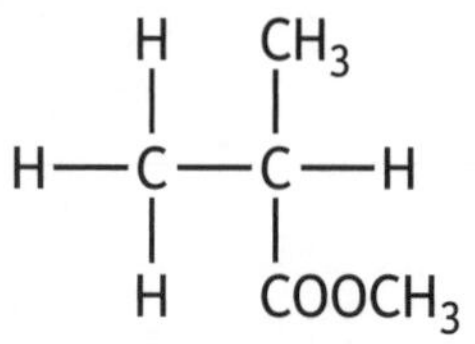

22. Hydrogen gas

    A   burns with a pop

    B   relights a glowing splint

    C   turns damp pH paper red

    D   turns limewater cloudy.

23. What is the charge on an iron ion in $Fe_2(SO_4)_3$?

    A   3–

    B   3+

    C   2–

    D   2+

24. Sodium sulfate solution reacts with barium chloride solution.

$$Na_2SO_4(aq) + BaCl_2(aq) \rightarrow BaSO_4(s) + 2NaCl(aq)$$

The spectator ions present in this reaction are

    A Ba^{2+} and Cl^-

    B Ba^{2+} and SO_4^{2-}

    C Na^+ and Cl^-

    D Na^+ and SO_4^{2-}

[Turn over

25. But-1-ene is a colourless, insoluble gas which is more dense than air but less dense than water.

Which of the following diagrams shows the most appropriate apparatus for collecting and measuring the volume of but-1-ene?

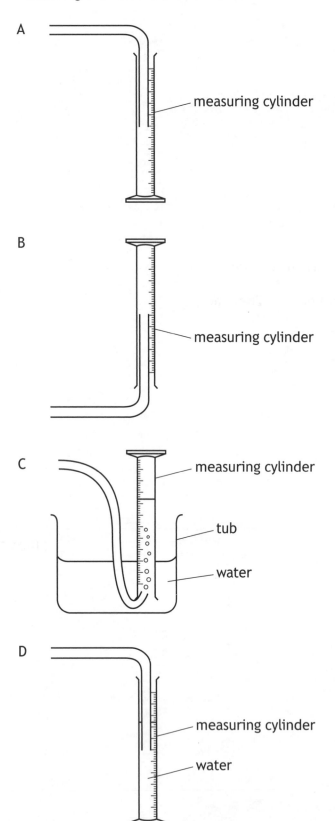

A   measuring cylinder

B   measuring cylinder

C   measuring cylinder
tub
water

D   measuring cylinder
water

[END OF SECTION 1. NOW ATTEMPT THE QUESTIONS IN SECTION 2 OF YOUR QUESTION AND ANSWER BOOKLET]

N5

Mark

National Qualifications
SPECIMEN ONLY

S813/75/01

Chemistry
Section 1—Answer Grid
And Section 2

Date — Not applicable

Duration — 2 hours 30 minutes

Fill in these boxes and read what is printed below.

Full name of centre

Town

Forename(s)

Surname

Number of seat

Date of birth

Day	Month	Year	Scottish candidate number

Total marks — 100

SECTION 1 — 25 marks

Attempt ALL questions.

Instructions for the completion of Section 1 are given on *Page two*.

SECTION 2 — 75 marks

Attempt ALL questions.

You may refer to the Chemistry Data Booklet for National 5.

Write your answers clearly in the spaces provided in this booklet. Additional space for answers and rough work is provided at the end of this booklet. If you use this space you must clearly identify the question number you are attempting. Any rough work must be written in this booklet. Score through your rough work when you have written your final copy.

Use **blue** or **black** ink.

Before leaving the examination room you must give this booklet to the Invigilator; if you do not, you may lose all the marks for this paper.

SECTION 1 — 25 marks

The questions for Section 1 are contained in the question paper S813/75/02.

Read these and record your answers on the answer grid on *Page three* opposite.

Use **blue** or **black** ink. Do NOT use gel pens or pencil.

1. The answer to each question is **either** A, B, C, or D. Decide what your answer is, then fill in the appropriate bubble (see sample question below).

2. There is **only one correct** answer to each question.

3. Any rough working should be done on the additional space for answers and rough work at the end of this booklet.

Sample Question

To show that the ink in a ball-pen consists of a mixture of dyes, the method of separation would be

    A   fractional distillation

    B   chromatography

    C   fractional crystallisation

    D   filtration.

The correct answer is **B** − chromatography. The answer **B** bubble has been clearly filled in (see below).

Changing an answer

If you decide to change your answer, cancel your first answer by putting a cross through it (see below) and fill in the answer you want. The answer below has been changed to **D**.

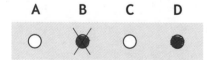

If you then decide to change back to an answer you have already scored out, put a tick (✓) to the **right** of the answer you want, as shown below:

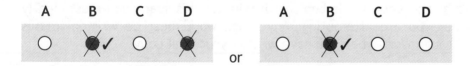

SECTION 1 — Answer Grid

	A	B	C	D
1	○	○	○	○
2	○	○	○	○
3	○	○	○	○
4	○	○	○	○
5	○	○	○	○
6	○	○	○	○
7	○	○	○	○
8	○	○	○	○
9	○	○	○	○
10	○	○	○	○
11	○	○	○	○
12	○	○	○	○
13	○	○	○	○
14	○	○	○	○
15	○	○	○	○
16	○	○	○	○
17	○	○	○	○
18	○	○	○	○
19	○	○	○	○
20	○	○	○	○
21	○	○	○	○
22	○	○	○	○
23	○	○	○	○
24	○	○	○	○
25	○	○	○	○

[BLANK PAGE]

DO NOT WRITE ON THIS PAGE

SECTION 2 — 75 marks

Attempt ALL questions

1. Graphs can be used to show the change in the rate of a reaction as the reaction proceeds.

   The graph shows the volume of gas produced in an experiment over a period of time.

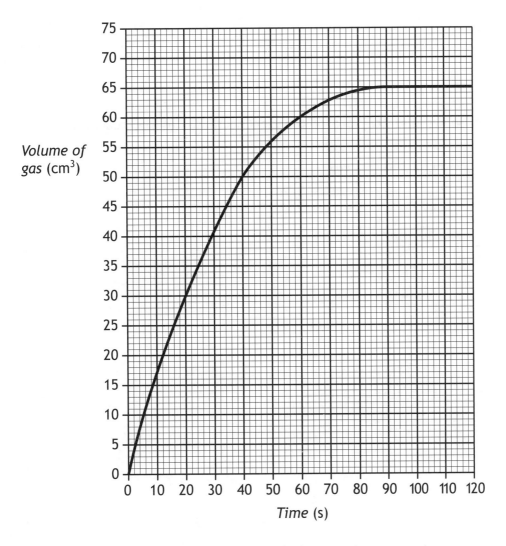

   (a) State the time, in seconds, at which the reaction stopped.　　　　1

[Turn over

MARKS | DO NOT WRITE IN THIS MARGIN

1. (continued)

(b) Calculate the average rate of reaction for the first 20 seconds.   **3**

Your answer must include the appropriate unit.

Show your working clearly.

(c) The graph shows that the rate of reaction decreases as the reaction proceeds.

Suggest a reason for this decrease.   **1**

MARKS | DO NOT WRITE IN THIS MARGIN

2. The group 7 element bromine was discovered by Balard in 1826.

Bromine gets its name from the Greek "bromos" meaning stench.

A sample of bromine consists of a mixture of two isotopes, $^{79}_{35}Br$ and $^{81}_{35}Br$.

(a) State what is meant by the term isotope.

1

(b) Complete the table for $^{79}_{35}Br$.

1

Isotope	Number of protons	Number of neutrons
$^{79}_{35}Br$		

(c) The sample of bromine has an average atomic mass of 80.

Suggest what this indicates about the amount of each isotope in this sample.

1

[Turn over

MARKS | DO NOT WRITE IN THIS MARGIN

2. (continued)

(d) In 1825 bromine had been isolated from sea water by Liebig who mistakenly thought it was a compound of iodine and chlorine.

Using your knowledge of chemistry, comment on why Liebig might have made this mistake.

3

MARKS | DO NOT WRITE IN THIS MARGIN

3. Antacid tablets are used to treat indigestion which is caused by excess acid in the stomach.

   Different brands of tablets contain different active ingredients.

Name of active ingredient	magnesium carbonate	calcium carbonate	magnesium hydroxide	aluminium hydroxide
Reaction with acid	fizzes	fizzes	does not fizz	does not fizz
Cost per gram (pence)	16	11	7·5	22
Mass of solid needed to neutralise $20\,cm^3$ of acid (g)	0·7	1·2	0·6	0·4
Cost of neutralising $20\,cm^3$ of acid (pence)		13·2	4·5	8·8

(a) Write the formula, showing the charge on each ion, for aluminium hydroxide.   1

(b) (i) Complete the table to show the cost of using magnesium carbonate to neutralise $20\,cm^3$ of acid.   1

   (ii) Using information from the table, state which **one** of the four active ingredients **you** would use to neutralise excess stomach acid.

   Explain your choice.   1

[Turn over

MARKS | DO NOT WRITE IN THIS MARGIN

4. Sulfur dioxide gas is produced when fossil fuels containing sulfur are burned.

   (a) When sulfur dioxide dissolves in water in the atmosphere "acid rain" is produced.

   Circle the correct phrase to complete the sentence.   1

   Compared with pure water, acid rain contains $\left\{\begin{array}{l}\text{a higher}\\\text{a lower}\\\text{the same}\end{array}\right\}$ concentration of hydrogen ions.

   (b) The table gives information about the solubility of sulfur dioxide.

Temperature (°C)	18	24	30	36	42	48
Solubility (g/100 cm³)	11·2	9·2	7·8	6·5	5·5	4·7

   (i) Draw a graph of solubility against temperature.

   Use appropriate scales to fill most of the graph paper.   4

   (Additional graph paper, if required, can be found on *Page twenty-eight.*)

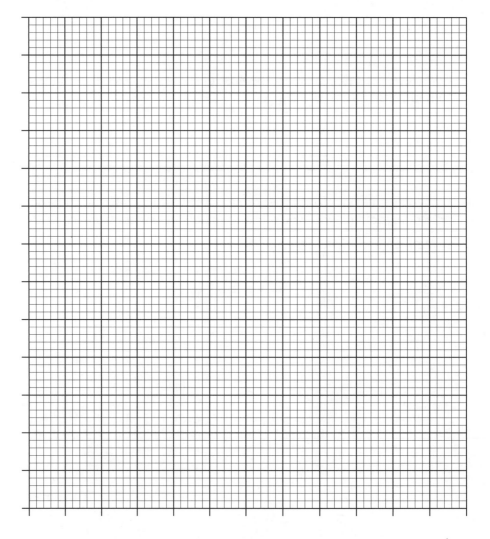

MARKS DO NOT WRITE IN THIS MARGIN

4. (b) (continued)

(ii) Estimate the solubility of sulfur dioxide, in g/100 cm^3, at 21 °C. 1

[Turn over

MARKS | DO NOT WRITE IN THIS MARGIN

5. A student investigated the reaction of carbonates with dilute hydrochloric acid.

(a) In one reaction lithium carbonate reacted with dilute hydrochloric acid. The equation for the reaction is:

$$Li_2CO_3(s) + HCl(aq) \rightarrow LiCl(aq) + CO_2(g) + H_2O(\ell)$$

   (i) Balance this equation. **1**

   (ii) Identify the salt produced in this reaction. **1**

(b) In another reaction 1·0 g of calcium carbonate reacted with excess dilute hydrochloric acid.

$$CaCO_3(s) + 2HCl(aq) \rightarrow CaCl_2(aq) + CO_2(g) + H_2O(\ell)$$

   (i) Calculate the mass, in grams, of carbon dioxide produced. **3**
     Show your working clearly.

MARKS | DO NOT WRITE IN THIS MARGIN

5. (b) (continued)

(ii) The student considered two methods to confirm the mass of carbon dioxide gas produced in this reaction.

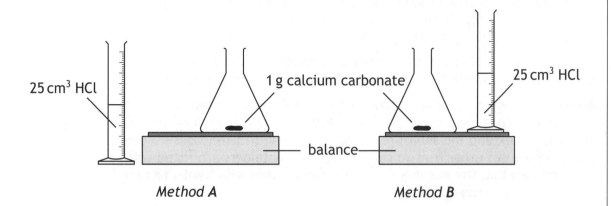

Method A   *Method B*

Method A	Method B
1. Add the acid from the measuring cylinder to the calcium carbonate in the flask.	1. Weigh the flask with the calcium carbonate and the acid in the measuring cylinder together.
2. Weigh the flask and contents.	2. Add the acid from the measuring cylinder to the calcium carbonate in the flask and replace the empty measuring cylinder on the balance.
3. Leave until no more bubbles are produced.	3. Leave until no more bubbles are produced.
4. Reweigh the flask and contents.	4. Reweigh the flask, contents and the empty measuring cylinder together.

Explain which method would give a more reliable estimate of the mass of carbon dioxide produced during the reaction.   2

[Turn over

MARKS | DO NOT WRITE IN THIS MARGIN

6. Read the passage below and answer the questions that follow.

Potassium Permanganate ($KMnO_4$)

Potassium permanganate's strong oxidising properties make it an effective disinfectant. Complaints such as athlete's foot and some fungal infections are treated by bathing the affected area in $KMnO_4$ solution.

In warm climates vegetables are washed in $KMnO_4$ to kill bacteria such as *E. coli*. Chemists use $KMnO_4$ in the manufacture of saccharin and benzoic acid.

Baeyer's reagent is an alkaline solution of $KMnO_4$ and is used to detect unsaturated organic compounds. The reaction of $KMnO_4$ with alkenes is also used to extend the shelf life of fruit. Ripening fruit releases ethene gas which causes other fruit to ripen. Shipping containers are fitted with gas scrubbers that use alumina or zeolite impregnated with $KMnO_4$ to stop the fruit ripening too quickly.

$$C_2H_4 \ + \ 4KMnO_4 \ \rightarrow \ 4MnO_2 \ + \ 4KOH \ + \ 2CO_2$$

Adapted from an article by Simon Cotton on "Soundbite molecules" in "Education in Chemistry" November 2009.

(a) Suggest an experimental test, including the result, to show that potassium is present in potassium permanganate.

You may wish to use the data booklet to help you.

1

(b) Suggest a pH for Baeyer's reagent.

1

(c) Name the gas removed by the scrubbers.

1

(d) Name a chemical mentioned in the passage which contains the following functional group.

1

(e) Zeolite is a substance that contains aluminium silicate.

Name the elements present in aluminium silicate.

1

MARKS | DO NOT WRITE IN THIS MARGIN

7. In the 2012 London Olympics, alkanes were used as fuels for the Olympic flame.

   (a) The torches that carried the Olympic flame across Britain burned a mixture of propane and butane.

   Propane and butane are members of the same homologous series.

   State what is meant by the term homologous series.   **1**

   (b) Natural gas, which is mainly methane, was used to fuel the flame in the Olympic cauldron.

      (i) Draw a diagram to show how **all** the outer electrons are arranged in a molecule of methane, CH_4.   **1**

      (ii) Methane is a covalent molecular substance. It has a low boiling point and is a gas at room temperature.

      Explain why methane is a gas at room temperature.   **2**

[Turn over

MARKS | DO NOT WRITE IN THIS MARGIN

8. Car manufacturers have developed vehicles that use ethanol as fuel.

   (a) The structure of ethanol is shown below.

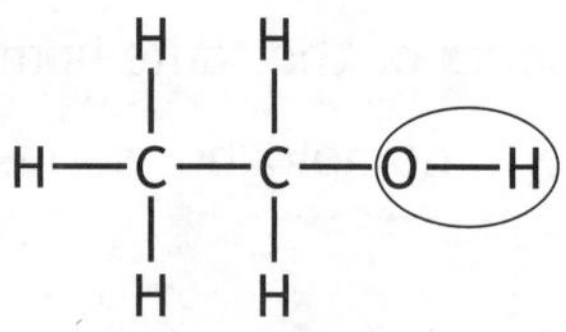

   Name the functional group circled in the diagram. **1**

   (b) Name the two substances produced when ethanol burns in a plentiful supply of oxygen. **1**

   (c) Ethanol can be produced from ethene as shown.

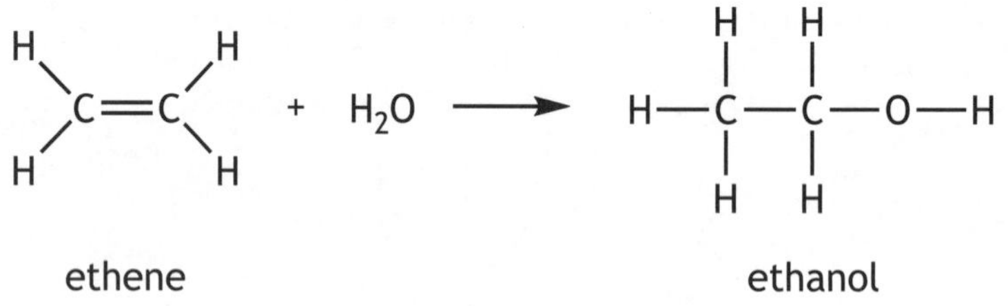

   ethene   ethanol

   (i) Name the **type** of chemical reaction taking place. **1**

MARKS | DO NOT WRITE IN THIS MARGIN

8. **(c)** **(continued)**

(ii) Draw a structural formula for a product of the following reaction.   **1**

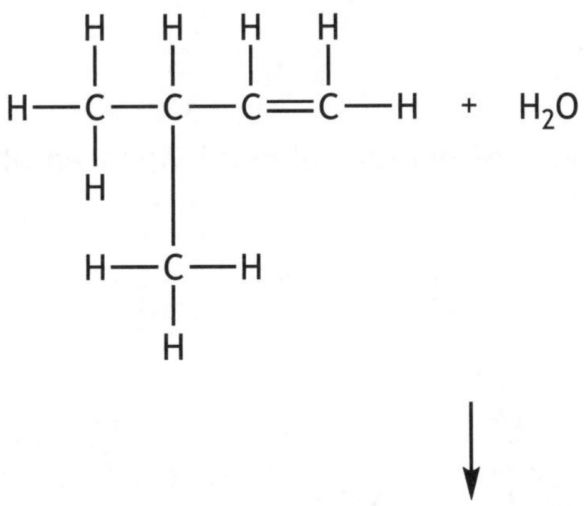

(d) Ethanol can be used to produce ethanoic acid.

(i) Draw a structural formula for ethanoic acid.   **1**

(ii) Name the family of compounds to which ethanoic acid belongs.   **1**

[Turn over

MARKS | DO NOT WRITE IN THIS MARGIN

9. Alkanes burn, releasing heat energy.

(a) State the term used to describe all chemical reactions that release heat energy.

1

(b) A student investigated the amount of energy released when an alkane burns using the apparatus shown.

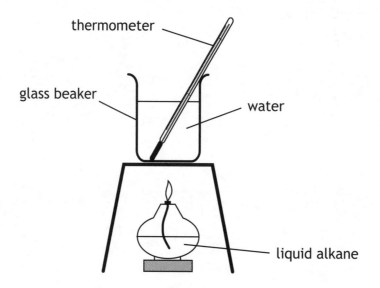

The student recorded the following data.

Mass of alkane burned	1 g
Volume of water	200 cm^3
Initial temperature of water	15 °C
Final temperature of water	55 °C

(i) Calculate the energy released, in kJ.

Show your working clearly.

3

MARKS | DO NOT WRITE IN THIS MARGIN

9. **(b)** **(continued)**

(ii) Suggest **one** improvement to the student's investigation.   1

(c) The table gives information about the amount of energy released when one mole of some alkanes are burned.

Name of alkane	Energy released when one mole of alkane is burned (kJ)
methane	891
ethane	1561
propane	2219
butane	2878

(i) Write a statement linking the amount of energy released to the number of carbon atoms in the alkane molecule.   1

(ii) Predict the amount of heat released, in kJ, when one mole of pentane is burned.   1

[Turn over

MARKS | DO NOT WRITE IN THIS MARGIN

10. Essential oils can be extracted from plants and used in perfumes and food flavourings.

(a) Essential oils contain compounds made up of a number of isoprene molecules joined together.

The shortened structural formula for isoprene is $CH_2C(CH_3)CHCH_2$.

Draw the **full** structural formula for isoprene.   **1**

(b) Essential oils can be extracted from the zest of lemons in the laboratory by steam distillation.

The process involves heating up water in a boiling tube until it boils. The steam produced then passes over the lemon zest which is separated from the water by glass wool. As the steam passes over the lemon zest it carries essential oils into the delivery tube. The condensed liquids (essential oils and water) are collected in a test tube placed in a cold water bath.

Complete the diagram to show the apparatus needed to collect the essential oils.   **1**

(An additional diagram, if required, can be found on *Page twenty-nine.*)

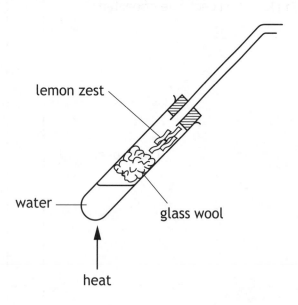

lemon zest

water

glass wool

heat

MARKS | DO NOT WRITE IN THIS MARGIN

10.   (continued)

(c)   Limonene, $C_{10}H_{16}$, is a compound found in lemon zest.

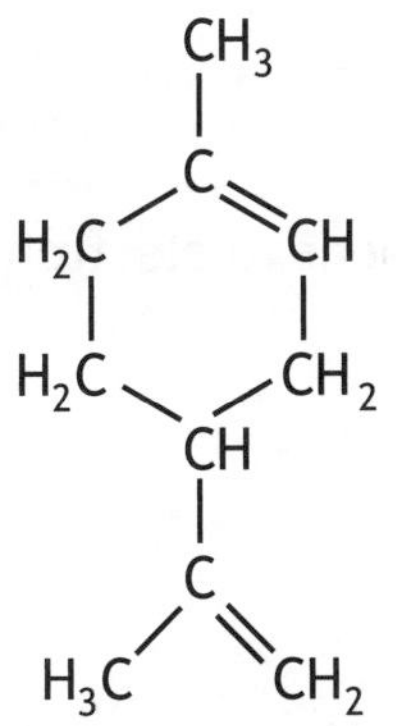

Write the molecular formula for the product formed when limonene reacts completely with bromine solution.

1

[Turn over

MARKS | DO NOT WRITE IN THIS MARGIN

11. Metals can be extracted from metal compounds by electrolysis.

(a) During electrolysis, metal ions are changed to metal atoms.

Name this type of chemical reaction.   1

(b) A student set up the following experiment to electrolyse copper(II) chloride solution.

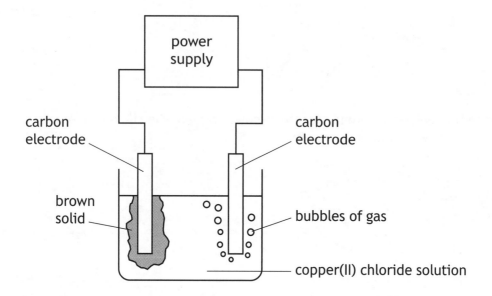

(i) Name the type of power supply that **must** be used to electrolyse the solution.   1

(ii) Complete the table by adding the charge on each electrode.   1

Observation at the _____ electrode	Observation at the _____ electrode
brown solid formed	bubbles of gas

MARKS | DO NOT WRITE IN THIS MARGIN

12. Urea, H_2NCONH_2, can be used as a fertiliser.

(a) Calculate the percentage of nitrogen in urea.   3

(b) Other nitrogen based fertilisers can be produced from ammonia.

Ammonia is produced in an industrial process using a catalyst.

$$N_2(g) \ + \ 3H_2(g) \ \rightleftharpoons \ 2NH_3(g)$$

(i) Name the industrial process that produces ammonia.   1

(ii) Suggest why a catalyst may be used in an industrial process.   1

(c) In another industrial process, ammonia is used to produce nitric acid.

Name the catalyst used in this process.   1

[Turn over

13. Vitamin C is found in fruits and vegetables.

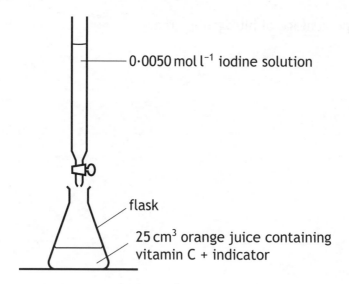

0·0050 mol l^{-1} iodine solution

flask

25 cm^3 orange juice containing vitamin C + indicator

Using iodine solution, a student carried out experiments to determine the concentration of vitamin C in orange juice.

The results of the experiments are shown.

Experiment	Initial volume of iodine solution (cm^3)	Final volume of iodine solution (cm^3)	Volume of iodine solution added (cm^3)
1	1·2	18·0	16·8
2	18·0	33·9	15·9
3	0·5	16·6	16·1

(a) (i) Name the piece of apparatus used to measure the volume of iodine solution added to the orange juice.

1

(ii) Calculate the average volume, in cm^3, of iodine solution that should be used in calculating the concentration of vitamin C.

Show your working clearly.

1

(b) Name the experimental method, carried out by the student, to accurately determine the concentration of vitamin C in the orange juice.

1

MARKS | DO NOT WRITE IN THIS MARGIN

14. In medicine, technetium-99m is injected into the body to detect damage to heart tissue.

It is a gamma-emitting radioisotope with a half-life of 6 hours.

(a) A sample of technetium-99m has a mass of 2 g.

Calculate the mass, in grams, of technetium-99m that would be left after 12 hours.

2

Show your working clearly.

(b) Suggest one reason why technetium-99m can be used safely in this way.

1

(c) Technetium-99m is formed when molybdenum-99 decays.

The decay equation is:

$$^{99}_{42}\text{Mo} \rightarrow {}^{99}_{43}\text{Tc} + X$$

Identify **X**.

1

[Turn over

MARKS | DO NOT WRITE IN THIS MARGIN

15. The concentration of chloride ions in water affects the ability of some plants to grow.

A student investigated the concentration of chloride ions in the water at various points along the river Tay.

The concentration of chloride ions in water can be determined by reacting the chloride ions with silver ions.

$$Ag^+(aq) + Cl^-(aq) \rightarrow AgCl(s)$$

A $20\,cm^3$ water sample gave a precipitate of silver chloride with a mass of $1\cdot435\,g$.

(a) Calculate the number of moles of silver chloride, AgCl, present in this sample.   **2**

(b) Using your answer to part (a), calculate the concentration, in mol l^{-1}, of chloride ions in this sample.   **2**

16. Nitrogen, phosphorus and potassium are elements essential for plant growth.

A student was asked to prepare a dry sample of a compound which contained **two** of these elements.

The student was given access to laboratory equipment and the following chemicals.

Chemical	Formula
ammonium hydroxide	NH_4OH
magnesium nitrate	$Mg(NO_3)_2$
nitric acid	HNO_3
phosphoric acid	H_3PO_4
potassium carbonate	K_2CO_3
potassium hydroxide	KOH
sodium hydroxide	$NaOH$
sulfuric acid	H_2SO_4
water	H_2O

Using your knowledge of chemistry, comment on how the student could prepare their dry sample.

3

[END OF SPECIMEN QUESTION PAPER]

MARKS | DO NOT WRITE IN THIS MARGIN

ADDITIONAL SPACE FOR ANSWERS

Additional graph paper for Question 4 (b) (i)

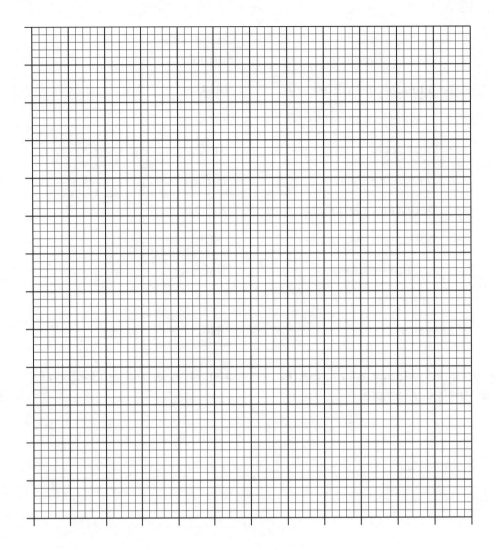

MARKS | DO NOT WRITE IN THIS MARGIN

ADDITIONAL SPACE FOR ANSWERS

Additional diagram for Question 10 (b)

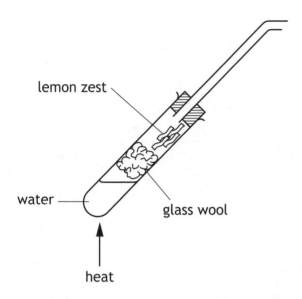

MARKS | DO NOT WRITE IN THIS MARGIN

ADDITIONAL SPACE FOR ANSWERS AND ROUGH WORK

NATIONAL 5

Answers

NATIONAL 5 CHEMISTRY 2016

Section 1

Question	Response
1.	D
2.	C
3.	D
4.	B
5.	B
6.	C
7.	A
8.	C
9.	B
10.	C
11.	B
12.	B
13.	A
14.	C
15.	D
16.	D
17.	A
18.	A
19.	C
20.	B

Section 2

1. (a) (i)

In the Nucleus		
Particle	Relative Mass	Charge
Proton	1	
Neutron		0 neutral no charge

Both required for 1 mark

(ii)

Outside the Nucleus		
Particle	Relative Mass	Charge
ELECTRON		- -1 negative

Both required for 1 mark

(b) 14.5

(c) (i) Pyramidal
   or
   trigonal pyramidal

   (ii) Haber

2. (a)

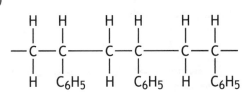

(b)

$$C=C$$

with H, CN on top and H, COOCH₃ on bottom

3. (a) Exothermic
   or
   exothermal

   (b) (i) 0·85 (with no working)   2
      Marks are awarded as follows:

      $\frac{29-12}{30-10}$ or $\frac{12-29}{10-30}$

      or

      $\frac{17}{20}$ (1)

      0·85   (1)

      (ii) Any value greater than 50 and less than or equal to 60

      (iii) Faster/quicker/increase/speed up

   (c) (i) Al(NO₃)₃ circled, underlined, etc.
      (ii) 0·36 (with no working)   2
      Marks are awarded as follows:
      0·01 moles gives 0·015 moles   (1)
      0·015 × 24 = 0·36   (1)
      This step on its own for 2 marks

4. (a) (i) Andalusite and kyanite

      Both required for 1 mark

      (ii)

Temperature (°C)	490–510
Pressure (kbar)	3·9–4·1

      Both required for 1 mark

   (b) 17 **or** 17·28 **or** 17·3 (with no working)   3
      With working:
      GFM 162   (1)
      28/162 × 100 (concept mark)   (1)
      This step on its own for 2 marks
      Correct arithmetic   (1)
      (This mark can only be awarded if the concept mark has been awarded.)

5. (a) Unreactive
   or
   does not react with water/air/alkalis/(almost all) acids
   or
   can be beaten into shape
   or
   found uncombined

(b) 118

(c) (i) $CO + O_2 \longrightarrow CO_2$ **All correct for 1 mark**

or

$CO + O_2 \xrightarrow{\text{Au}} CO_2$

(ii) Catalyst **or** catalysis

or

speeds up the reaction

or

allows less energy/heat to be used for the reaction

or

lowers activation energy

(d) (hydroxide) (hydroxide)
(hydrogen) (hydrogen)

Both required for 1 mark

6. (a) (i) Flame test or correct description e.g. burn it/ fertiliser/potassium, put in Bunsen flame, etc.
and
purple/lilac

Both required for 1 mark

(ii) To add/provide/supply nitrogen or it contains nitrogen. (Any wording that implies that nitrogen needs nitrogen.)

(b) $H_3PO_4 + 3NH_4OH \longrightarrow (NH_4)_3PO_4 + 3H_2O$

7. (a) This is an open-ended question.
1 mark: The student has demonstrated a limited understanding of the chemistry involved. The candidate has made some statement(s) which is/are relevant to the situation, showing that at least a little of the chemistry within the problem is understood.
2 marks: The student has demonstrated a reasonable understanding of the chemistry involved. The student makes some statement(s) which is/are relevant to the situation, showing that the problem is understood.
3 marks: The maximum available mark would be awarded to a student who has demonstrated a good understanding of the chemistry involved. The student shows a good comprehension of the chemistry of the situation and has provided a logically correct answer to the question posed. This type of response might include a statement of the principles involved, a relationship or an equation, and the application of these to respond to the problem. This does not mean the answer has to be what might be termed an "excellent" answer or a "complete" one.

8. (a) The correct structural formula for isoprene e.g.

or

(However if CH_3 is used the bond must be going to the carbon.)

(b) Diagram showing delivery tube passing into a test tube which is placed in a water/ice bath. Delivery tube must extend close enough to the neck of the test tube to ensure the vapour can enter the test tube.

(c) (i) Addition/additional
or
bromination

(ii) $C_{10}H_{16}Br_4$

9. (a) **Method A**   1
For explanation of accuracy of A
(or inaccuracy of B) based on   1
• heat loss
• heat transfer
• mass loss (due to ethanol being combusted/ used up)
e.g.
• **Method A** because more heat is transferred to water
• **Method B** because less heat is transferred to water
• **B** releases more heat to the surroundings
(*The second mark for the explanation will **not** be awarded if the first mark is not gained.*)

(b) (i) If the alcohol is 2-ol then less energy is released compared with 1-ol or vice versa.
or
As you move from one to two (carbon/position) then the energy decreases or vice versa.
or
As it (the position of the functional group) increases/gets higher, the energy released decreases or vice versa.
or
Functional group (or it/hydroxyl/-OH) on (position) 1/end carbon/first carbon – energy released is greater/higher/bigger/increases.
or
Functional group (or it/hydroxyl/-OH) on position 2/not on the end carbon energy released is smaller/lower/decreases.
or
As it/functional group goes further along/ further down/further up the lower the energy or vice versa.

(ii) 3967–3971

(c) 55 or 55·02 (with no working)   3
With working:
Use of correct concept
$\Delta T = Eh/cm$
with **both 4·18 and 23** correctly substituted   (1)
0·1 (with or without concept)   (1)
Correct arithmetic   (1)
(This mark can only be awarded if the concept mark has been awarded.)

10. (a) Electrolyte

(b) (i) From zinc to copper on or near the wire/ voltmeter

(ii) Ion bridge
or
salt bridge

(c) (i) Oxidation

(ii) $Br_2(\ell) + SO_3^{2-}(aq) + H_2O(\ell) \longrightarrow 2Br^-(aq) + SO_4^{2-}(aq) + 2H^+(aq)$

or

$$Br_2(\ell) + SO_3^{2-}(aq) + H_2O(\ell)$$

$$\downarrow$$

$$2Br^-(aq) + SO_4^{2-}(aq) + 2H^+(aq)$$

(iii) Carbon or graphite

11. (a) Methoxypropane (spelling must be correct)

(b) $C_nH_{2n+2}O$

or

$C_nH_{2n+2}O_1$

or

$C_nH_{n2+2}O$

(c)

weak

STRONG

WEAK

strong

Both required for 1 mark

(d) (i) Ethene/Eth-1-ene

or

ethylene

(ii) Any acceptable full, shortened or abbreviated structural formula e.g.

$$H-\underset{\underset{H}{|}}{\overset{\overset{H}{|}}{C}}-\underset{\underset{H}{|}}{\overset{\overset{H}{|}}{C}}\overset{O\diagdown\diagup C}{}\overset{H \quad H}{}H$$

$$H-C-C<^{O}_{}>C<^{CH_3}_{H}$$

$$H_2C-CH<^{O}>C<^{H\ \ H}$$

$$H_2C-CH<^{O}>CH_3$$

12. (a) Hydroxyl

(b) Weak acid, strong base/alkali

Both required for 1 mark

(c) 15·0

(d) 0·02

13. This is an open-ended question.

1 mark: The student has demonstrated a limited understanding of the chemistry involved. The candidate has made some statement(s) which is/are relevant to the situation, showing that at least a little of the chemistry within the problem is understood.

2 marks: The student has demonstrated a reasonable understanding of the chemistry involved. The student makes some statement(s) which is/are relevant to the situation, showing that the problem is understood.

3 marks: The maximum available mark would be awarded to a student who has demonstrated a good understanding of the chemistry involved. The student shows a good comprehension of the chemistry of the situation and has provided a logically correct answer to the question posed. This type of response might include a statement of the principles involved, a relationship or an equation, and the application of these to respond to the problem. This does not mean the answer has to be what might be termed an "excellent" answer or a "complete" one.

NATIONAL 5 CHEMISTRY 2017

Section 1

Question	Response
1.	D
2.	C
3.	A
4.	D
5.	A
6.	B
7.	C
8.	B
9.	A
10.	A
11.	D
12.	C
13.	C
14.	B
15.	D
16.	C
17.	A
18.	B
19.	A
20.	C

Section 2

1. (a) Isotope(s)

(b) Different numbers of neutrons

or

the atoms have 18, 20 or 22 neutrons

(c) 36

or

$^{36}_{18}Ar$

or

^{36}Ar

2. (a) Carbon nanotube

or

Nanotube

(b) Lithium or Li

(c) 20·5 with no working
21 with correct working

Partial marking:
Demonstration of the correct use of the
relationship concept.
ie 41/2. (1)
or
41/1 = 41 (1)
Working must be shown

3. (a) Diagram showing two chlorine atoms with one
pair of bonding electrons; must show **all outer**
electrons e.g.

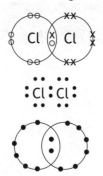

(b) Tetrahedral/tetrahedron

(c) Gains an electron (from sodium)

or

Indication that the electron arrangement
increases by 1

eg electron arrangement goes from 2.8.7 to 2.8.8,
outer electron number goes from 7 to 8.

(d) Low, no

Both must be correct for 1 mark

High, no

Both must be correct for 1 mark

4. (a) (i)

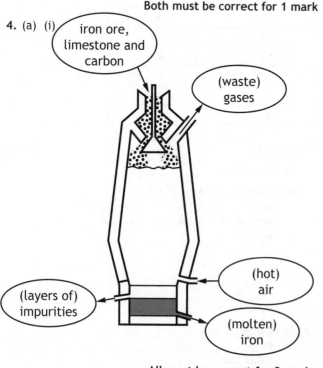

All must be correct for 2 marks
2/3 must be correct for 1 mark
0/1 correct for 0 marks

(ii) Iron would not melt/be molten/liquid or able to
flow
or
Iron would be solid
or
Iron needs to be molten/liquid/flowing

(b) $Fe^{2+} \rightarrow Fe^{3+} + e^-$
or

$Fe^{2+} - e \rightarrow Fe^{3+}$

State symbols are not required;
however, if given they must be correct

5. (a) (i) 14 (days)
No units required but no mark is awarded if
wrong unit is given. Wrong units would only
be penalised once in the paper.

(ii) 42 (days)
No units required but maximum of 1 mark is
awarded if wrong unit is given

Partial marking:
3 half-lives (1)
or
Correct number of days for an incorrect
number of half-lives (1)
Working must be shown

(b) beta
or
β
or
$_{-1}^{0}\beta$
or
$_{-1}^{0}e$
or
$_{-1}^{0}e^-$

6. This is an open-ended question.
1 mark: The student has demonstrated a limited
understanding of the chemistry involved. The candidate
has made some statement(s) which is/are relevant
to the situation, showing that at least a little of the
chemistry within the problem is understood.
2 marks: The student has demonstrated a reasonable
understanding of the chemistry involved. The student
makes some statement(s) which is/are relevant to the
situation, showing that the problem is understood.
3 marks: The maximum available mark would be
awarded to a student who has demonstrated a good
understanding of the chemistry involved. The student
shows a good comprehension of the chemistry of the
situation and has provided a logically correct answer to
the question posed. This type of response might include
a statement of the principles involved, a relationship
or an equation, and the application of these to respond
to the problem. This does not mean the answer has to
be what might be termed an "excellent" answer or a
"complete" one.

7. (a) Carboxyl
or

O
‖
C
\
OH

or
COOH

(b) (i) Any acceptable structural formula for butanoic acid
e.g.
CH$_3$CH$_2$CH$_2$COOH
CH$_3$(CH$_2$)$_2$COOH

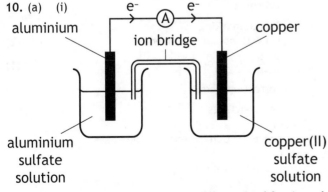

(ii) Butanoic acid or it has bigger/stronger/more forces (of attraction) **(1)**
Between molecules **or** mention of intermolecular attractions **(1)**
If neither of these two points are given, a maximum of 1 mark can be awarded for:
Butanoic acid or it is bigger/has more carbons or hydrogens or atoms/longer carbon chain

8. (a) (i) Glowed brighter/more brightly than zinc
or
Glowed most brightly/very brightly/white light

(ii) Faster/higher/speed up/increase

(b) Magnesium
or
Mg

9. (a) They have the same general formula
AND
similar/same <u>chemical</u> properties
Both required for 1 mark

(b) Isomer(s)

(c) (i) Increasing carbon chain length/number of carbons takes more time (longer, slower)
or
Decreasing carbon chain length/number of carbons takes less time (faster, quicker)
or
Straight chain takes more time (longer, slower) than branched chain
or
Branched chain takes less time (faster, quicker) than straight chain

(ii) Indication that the expected position occurs anywhere on the horizontal line between ethane and 2-methylpropane.

10. (a) (i)

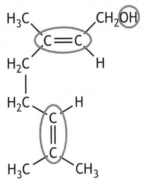

aluminium ...... copper
ion bridge
aluminium sulfate solution ...... copper(II) sulfate solution
All required for 1 mark

(ii) $3Cu^{2+} + 2Al \rightarrow 3Cu + 2Al^{3+}$
Accept correct multiples
Zero marks awarded for -electrons shown in equation, unless clearly scored out
State symbols are not required however if given they must be correct

(b) 15·79 or 15·8 or 16 (%)
Partial marking:
GFM = 342 **(1)**
54/342 × 100 (concept mark) **(1)**
This step on its own 2 marks
Calculation of final answer using the correct relationship **(1)**
No units required but maximum of 2 marks can be awarded if wrong unit is given

11. (a) pH of solution goes down/decreases/goes below 7/goes to a value less than 7 from 7 **because** the H$^+$ ion/hydrogen ion concentration increases/goes up or more H$^+$ than OH$^-$/H$^+$ > OH

Partial marking:
pH of solution goes down/decreases/goes below 7/goes to a value less than 7 from 7 **(1)**
or
H$^+$ ion/hydrogen ion concentration increases/goes up/more H$^+$ than OH$^-$/H$^+$ > OH$^-$ **(1)**

(b) Decreases/goes down/gets lower

12. (a) Select one of these functional groups for 1 mark:

H$_3$C—C=C—CH$_2$OH ... H$_2$C ... H ... H$_2$C ... C=C ... H$_3$C ... CH$_3$

(b) Ester(s)

(c) 21 (g)

Partial marking:
1 mark for **either**
Both GFMs
ie 154 and 210
or
Moles of geraniol
ie (15·4/154) = 0·1 mol
----

1 concept mark for **either**

$15·4 \times \dfrac{\text{GFM of ester}}{\text{GFM of geraniol}}$

ie 15.4 × (210/154)
or
Moles of geraniol × GFM of ester
ie 0·1 × 210
(Either of these two steps on their own with all correct substitutions 2 marks)
1 mark for calculated final answer provided the concept mark has been awarded
No units required but a maximum of two marks can be awarded if wrong unit is given.

13. (a) C$_n$H$_{2n-2}$
or
C$_n$H$_{n2-2}$
or
C$_n$H$_{2(n-1)}$

(b) (i)

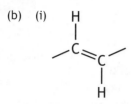

(ii) Addition/additional

(c) (i)

H—C≡C—C—H structure (ethyne-like structure with methyl groups, showing H—C—C≡C—C—H with H atoms)

(ii) The two bromine atoms are not next to one another.
or
The two bromines are separated by a hydrogen.
or
The two bromine branches are not next to one another.

14. (a) (i) Any correct shortened or full structural formula for hexan-1-ol

(ii) 188 (kJ)

   No units required but no mark is awarded if wrong unit is given.

(b) 3·9 or 3·91 or 4 (kJ kg^{-1} °C^{-1})
   No units required but a maximum of two marks can be awarded if wrong unit is given.

Partial marking:
Using the correct concept of
$c = E_h / m \Delta T$
with $E_h = 13·3$ (1)
0·1 **and** 34   (1)

A further mark can be awarded for the candidate's calculated answer **only** if the mark for the concept has been awarded.

Alternatively
— 13300 and 0·1 can be used but the final answer should be 3912 J kg^{-1} °C^{-1} (units must be shown and correct for 3 marks to be awarded). If no unit, or the unit given in question is used then 2 marks are awarded as the mark for the final calculated answer is not awarded.

Or alternatively
— the answer, 3912, can be divided by 1000 to give the correct answer in kJ kg^{-1} °C^{-1}.

Units must be shown and correct for 3 marks to be awarded). If no unit, or the unit given in question is used then 2 marks are awarded as the mark for the final calculated answer is not awarded.

Section 1

Question	Answer	Max mark
1.	B	1
2.	C	1
3.	A	1
4.	D	1
5.	A	1
6.	D	1
7.	A	1
8.	C	1
9.	C	1
10.	A	1
11.	D	1
12.	A	1
13.	C	1
14.	B	1
15.	A	1
16.	B	1
17.	C	1
18.	D	1
19.	D	1
20.	A	1
21.	B	1
22.	A	1
23.	B	1
24.	C	1
25.	C	1

Section 2

1. (a) 86—90 (seconds)

   (Units are not required, but 0 marks should be awarded for the correct answer if incorrect unit is given.)

   (b) 1·5 cm^3 s^{-1} 3
   Partial marking:
   1·5 with no unit/incorrect unit   (2)
   $\frac{30-0}{20-0}$ or $\frac{30}{20}$ or $\frac{0-30}{0-20}$ (1)
   Correct unit cm^3 s^{-1} (1)

   (c) Less reactants
   or
   concentration of reactants decreases
   or
   reactants are used up
   or
   less chance of particles colliding
   or
   equivalent answer

2. (a) Atoms with same atomic number/number of protons/positive particles but different mass number/number of neutrons

(b) Protons = 35
Neutrons = 44
Both required for 1 mark

(c) Equal amounts/proportions/abundance
or
same number of each
or 50:50
or equivalent answers

(d) This is an open ended question.
1 mark: The candidate has demonstrated a limited understanding of the chemistry involved. The candidate has made a/some statement(s) which is/are relevant to the situation, showing that at least a little of the chemistry within the problem is understood.
2 marks: The candidate has demonstrated a reasonable understanding of the chemistry involved. The candidate has made a/some statement(s) which is/are relevant to the situation, showing that the problem is understood.
3 marks: The candidate has demonstrated a good understanding of the chemistry involved. The candidate shows a good comprehension of the chemistry of the situation and has provided a logically correct answer to the question posed. This type of response might include a statement of the principles involved, a relationship or an equation, and the application of these to respond to the problem. This does not mean the answer has to be what might be termed an "excellent" answer or a "complete" one.

3. (a) $Al^{3+}(OH^-)_3$

(b) (i) 11·2 (pence).

Unit is not required, but 0 marks can be awarded for the correct answer if incorrect unit is given. Incorrect units would only be penalised once in any paper.

(ii) Named active ingredient with an appropriate reason.

e.g. magnesium hydroxide
  — cheapest/doesn't fizz

aluminium hydroxide — need to take least amount

4. (a) (a higher)

(b) (i) For appropriate format: scatter graph — i.e. a graph in which points are plotted with their x and y values representing temperature and solubility 1

The axis/axes of the graph has/have suitable scale(s). For the graph paper provided within the question paper, the selection of suitable scales will result in a graph that occupies at least half of the width and half of the height of the graph paper 1

The axes of the graph have suitable labels and units 1

All data points plotted accurately with a line of best fit drawn 1

(ii) 10·2 − 10·3 (g/100 cm³)
or
a value correctly read from candidate's graph (allow ½ box tolerance)

(Units are not required, but 0 marks can be awarded for correct answer if incorrect unit is given.)

5. (a) (i) $Li_2CO_3 + 2HCl \rightarrow 2LiCl + CO_2 + H_2O$
(or correct multiples)

(ii) LiCl
or
lithium chloride

(b) (i) 0·44 (g) 3

(Units are not required, but a maximum of 2 marks can be awarded for the correct answer if incorrect unit is given.)

Partial marking:
Both *GFMs* 100 and 44 (1)

Correct application of the relationship between moles and mass (1)

This could be shown:

• by working containing the two expressions

$$\frac{1}{candidate's\ GFM\ for\ CaCO_3}$$

and

no. moles CO_2 × candidate's GFM CO_2

or

• by working showing correct proportionality

$$1 \leftrightarrow \frac{candidate\ GFM\ CO_2}{candidate\ GFM\ CaCO_3}$$

Where the candidate has been awarded the mark for the correct application of the relationship between moles and mass, a further mark can be awarded for correct follow through to a final answer. (1)

(ii) Method B (1)

Gas is lost in method A before starting mass taken
or
gas is lost before all acid is added
or
no total mass of all reactants at the start of experiment
or
equivalent response (1)

6. (a) flame test (or correct description) **and** lilac/purple
Both required for 1 mark.

(b) greater than 7
or
any numerical value greater than seven

(c) Ethene
Accept correct formula.

(d) Benzoic acid

(e) Aluminium, silicon and oxygen
Accept correct formulae.

7. (a) group/family/chemicals/compounds with same general formula and same/similar chemical properties

    Both parts required for 1 mark.

   (b) (i) Diagram showing carbon with four hydrogen atoms: each of the four overlap areas must have two electrons in or on overlap area (cross, dot, petal diagram).

      e.g.

      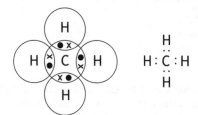

      (ii) Weak bond/attraction between molecules   1
                                                     1

8. (a) Hydroxyl

   (b) Carbon dioxide
      and
      water

      Both required for 1 mark.
      Accept correct formulae.

   (c) (i) Addition
      or
      hydration

      (ii)

      ```
          H   H   H   H
          |   |   |   |
      H — C — C — C — C — H
          |   |   |   |
          H   |   H   O — H
              |
          H — C — H
              |
              H
      ```

      or

      ```
          H   H   H       H
          |   |   |       |
      H — C — C — C ————— C — H
          |   |   |       |
          H   |   O — H   H
              |
          H — C — H
              |
              H
      ```

      Accept full or shortened structural formula.

   (d) (i)

      ```
          H       O
          |      //
      H — C — C
          |      \
          H       O — H
      ```

      Accept full or shortened structural formula.

      (ii) Carboxylic acid
         (Accept alkanoic acid.)

9. (a) Exothermic

   (b) (i) 33·44 (kJ) (3)

      Partial marking:
      Using $cm\Delta T$ with $c = 4·18$ (1)

      To be awarded this concept mark, candidates do not specifically need to write $cm\Delta T$. The concept mark is awarded for using this relationship with three values, one of which must be 4·18

      For values 0·2 (kg) and 40 (°C)   (1)

      A further mark can be awarded for arithmetical follow through to the candidate's answer only if the mark for the $cm\Delta T$ concept has been awarded.   (1)

      (Units are not required, but a maximum of 2 marks can be awarded for the correct answer if incorrect unit is given.)

      (ii) Draught insulation
         or
         use metal beaker
         or
         repeat to get average
         or
         any reasonable answer.

   (c) (i) As the number of carbons increases the energy released increases.
      or
      As the number of carbons decreases the energy released decreases.
      or
      The energy increases as the number of carbons increases.
      or
      The energy decreases as the number of carbons decreases.

      (ii) 3520 to 3550 (kJ)

      (Units are not required, 0 marks can be awarded for the correct answer if incorrect unit is given.)

10. (a)

   ```
                  H
                  |
              H — C — H
                  |
          H       |   H   H
          |       |   |   |
      H — C = C — C = C — H
   ```

   or

   ```
          H   CH₃  H   H
          |   |    |   |
      H — C = C — C = C — H
   ```

   In the formula above, the bond to the methyl group must be correctly aligned with the C atom of the group.

   (b) Diagram showing delivery tube passing into a test tube which is placed in a water/ice bath.

   Delivery tube must extend close enough to the neck of the test tube to ensure the vapour can enter the test tube.

   (c) $C_{10}H_{16}Br_4$

11. (a) Reduction

　(b) (i) d.c.

　　(ii) Negative — (brown solid formed)
　　　　Positive — (bubbles of gas)
　　　　　　　　Both required for one mark.

12. (a) 46·67/46·7/47　　　　　　　　　　**3**

　Partial marking:
　GFM = 60　　　　　　　　　　　　(1)

$$\frac{28}{\text{candidate's GFM}} \times 100 \qquad (1)$$

　Calculation of final answer using the relationship

$$\% \text{ by mass} = \frac{m}{GFM} \times 100 \qquad (1)$$

　(b) (i) Haber (-Bosch)

　　(ii) Speeds up reaction
　　　　or
　　　　Less energy/temperature/heat required

　(c) Platinum
　　(Accept platinum and rhodium (alloy).

13. (a) (i) Burette

　　(ii) 16 or 16·0 (cm^3)
　　　　(Units are not required; 0 marks can be awarded
　　　　for the correct answer if incorrect unit is given.)

　(b) Titration

14. (a) 0·5 (g)　　　　　　　　　　　　**2**
　　(Units are not required but a maximum of 1 mark
　　can be awarded for the correct answer if incorrect
　　unit is given.)

　Partial marking:
　1 mark can be awarded for either:
　• 2 half lives
　　or
　　mass correctly calculated for an incorrect number
　　of half-lives shown.

　(b) Short half-life
　　or
　　would not last long in the body
　　or
　　gamma would go right through body
　　or
　　equivalent response

　(c) beta/β/$_{-1}^{0}e$/$_{-1}^{0}\beta$

　　The charge on the beta particle does not need to be
　　shown.

　　Do not accept electron without atomic and mass
　　numbers, i.e. e or e-

15. (a) 0·01 (mol)　　　　　　　　　　　**2**
　　(Units are not required but a maximum of 1 mark
　　can be awarded for the correct answer if incorrect
　　unit is given.)

　Partial marking:
　1 mark can be awarded for either

　• 143·5 g

　or

　• correctly calculated answer for $\dfrac{1 \cdot 435}{\text{incorrect GFM}}$

　(b) 0·5 (mol l^{-1})　　　　　　　　　　**2**

　　(Units are not required but a maximum of 1 mark
　　can be awarded for the correct answer if incorrect
　　unit is given

　Partial marking:
　1 mark can be awarded for either

　• $\dfrac{0 \cdot 01}{0 \cdot 02}$

　or

　• correctly calculated answer for
　　$\dfrac{0 \cdot 01}{20}$

　　If correct relationship is used but volume not
　　converted to litres e.g. 0·01/20 maximum 1 mark.

16. This is an open ended question.
　1 mark: The candidate has demonstrated a limited
　understanding of the chemistry involved. The candidate
　has made a/some statement(s) which is/are relevant
　to the situation, showing that at least a little of the
　chemistry within the problem is understood.
　2 marks: The candidate has demonstrated a reasonable
　understanding of the chemistry involved. The candidate
　has made a/some statement(s) which is/are relevant to
　the situation, showing that the problem is understood.
　3 marks: The candidate has demonstrated a good
　understanding of the chemistry involved. The candidate
　shows a good comprehension of the chemistry of the
　situation and has provided a logically correct answer to
　the question posed. This type of response might include
　a statement of the principles involved, a relationship
　or an equation, and the application of these to respond
　to the problem. This does not mean the answer has to
　be what might be termed an "excellent" answer or a
　"complete" one.

Acknowledgements

Permission has been sought from all relevant copyright holders and Hodder Gibson is grateful for the use of the following:

The article 'Gold – a very useful metal' adapted from 'Education in Chemistry' Volume 45, Nov 2008 © Royal Society of Chemistry (2016 Section 2 page 14);
Image © Alexey Lysenko/Shutterstock.com (2016 Section 2 page 28);
An extract adapted from 'Hydrogen Storage' from InfoChem Magazine (RSC), Nov 2008 © Royal Society of Chemistry (2017 Section 2 page 6);
An extract adapted from 'Potassium Permanganate' by Simon Cotton from 'Education in Chemistry',
Nov 2009 © Royal Society of Chemistry (2017 SQP Section 2 page 14).